Additional Mathematics

Additional Mathematics

George N. Frempong

Copyright © 2020, 2015 by George N. Frempong

All rights reserved. No part of this book may be reproduced or transmitted in any form or any means, electronic or mechanical, including photocopying, recording, or by any information storage and retrieval system, without permission in writing from the copyright owner.

Contents

Preface

1. Sets — 1
2. Functions — 13
3. Quadratic Functions — 37
4. Indices and Surds — 59
5. Polynomials — 77
6. Simultaneous Equations — 85
7. Exponential and Logarithmic Functions — 91
8. Straight Line Graphs — 107
9. Circular Measure — 125
10. Trigonometry — 137
11. Permutations and Combinations — 159
12. Binomial Expansions — 171
13. Vectors — 179
14. Matrix Algebra — 207
15. Differentiation and Integration — 225

Answers to Exercises — 307

Preface

The book covers all the topics in the Cambridge O Level Additional Mathematics syllabus. The text has been updated and refined to improve clarity of the presentations. The clarity of the text makes the book excellent for self-tuition.

The purpose of this book is to provide you with the tools you need to be successful in your Additional Mathematics course. The exercises at the end of each section are varied, providing basic skills and understanding of concepts. Every chapter ends with a review exercise. The review exercises give you the opportunity to see where you need additional help or practice.

It is impossible to look back and recall the origins of most of the materials in this book. I am very grateful to all whose work I may have benefitted from.

1 Sets

1.1 Set and Set Notation

A well-defined collection of objects is called a set. Examples include a collection of books, a class of students and group of mathematics teachers.

Each member of a set is called an element. All elements of a set share a common property amongst them.

Set Notation

A set is described by listing its element between curly brackets. For instance, the set whose elements are 2, 4, 6 and 8 is written as {2, 4, 6, 8}.

Naming Sets

Capital letters are used to denote sets, and elements of sets are denoted in small letters. For example, we can denote the set {a, b, c, d, e} by the letter A, and write A = {a, b, c, d, e}.

Indicating Membership of a Set

The symbol $\in$ is used to indicate that an element belongs to a set. For example, if A = {2, 4, 5, 6, 7, 8}, then $6 \in A$. The symbol $\notin$ is used to indicate that an element is not an element of a set. For example, $9 \notin A$.

Set-Builder Notation

Sometimes, we denote sets by stating properties common to their elements. For example, the set of real numbers between 1 and 2, can be denoted by $\{x: x \in R, \ 1 < x < 2\}$. The colon " : " means " such that ". The vertical bar " | " can also be used.

Finite and Infinite Sets

A set is said to be finite if it is either empty or its element can be counted. Otherwise the set is said to be infinite. For example, the set $\{-2, -1, 0, 2, 4\}$ is finite, and the set {1, 2, 3, 4, 5, 6 ...} is infinite. The three dots indicate that the elements in the set continue in the same pattern.

The Number of Elements in a Set

The number of elements in a set A is denoted by $n(A)$. For example, if A = {10, 11, 12, 13, 15}, then $n(A) = 5$.

Equal Sets

Two sets A and B are equal, written A = B, if they contain exactly the same elements.

Empty Set

A set that has no elements is called an empty set (or null set). The empty set is denoted by Ø or { }. It is important to note that the empty set is not written as {Ø}.

Universal Sets

A set that contains all the elements for a specific discussion is called the universal set. The universal set is denoted by ξ or $\cup$.

Subsets

Set A is said to be a subset of set B when all elements of set A are also elements of set B, written A ⊆ B. For example, the set of natural numbers is a subset of the set of integers. If set A is not a subset of set B, we write A ⊄ B.

Proper Subsets

Set A is a proper subset of set B, denoted by A ⊂ B, when all elements of set A are also elements of set B, and there is at least one element of set B which is not an element of set A. For example, if A = {2, 5, 6} and B = {1, 2, 5, 6, 8}, then A ⊂ B. If set A is not a proper subset of B, we write A ⊄ B.

Note the following.

1. An empty set is a subset of every set.

2. Every set is a subset of itself.

3. A = B if and only if A ⊂ B and B ⊂ A.

Set and Set Notation

The Number of Subsets

A set with n elements has 2^n subsets. For example, a set containing 5 elements has 2^5, i.e. 32 subsets.

Exercise 1.1

List the elements of each set between curly brackets.

1. The set of whole numbers 5 through 12.

2. The set of multiples of 6 between 10 and 18.

3. The set of odd numbers between 15 and 30.

4. $\{x : x \text{ is an integer between } -5 \text{ and } 4\}$.

5. $\{x : x \text{ is a square of } -3\}$.

Classify each statement as true or false.

6. $2 \in \{x: x \text{ is a real number}\}$.   7. $9 \in \{x: x \text{ is a prime number}\}$.

8. $6 \in \{5, 6, 7\}$ 9. $7 \notin \{2, 4, 5\}$

10. $2 \in \{\text{prime numbers}\}$ 11. $8 \notin \{7, 8, 9\}$

12. $-3 \in \{x: x \text{ is a real number}\}$

Classify each set as finite or infinite.

13. $\{x: x \text{ is a multiple of } 2\}$

14. $\{x: 3x + 2 = 8\}$ 15. $\{x: x \in R, 5 < x < 6\}$

16. $\{x: x \text{ is a factor of } 20\}$ 17. $\{x: x + 3 > 0\}$

18. $\{x: x \in N, 10 < x < 90\}$

Find the number of elements of each set.

19. $\{x: x \text{ is a factor of } 10\}$ 20. $\{x: x \text{ is an even prime number}\}$

21. $\{x: x \text{ is an integer between } -8 \text{ and } 5\}$

22. $\{x: x \text{ is a multiply of } 3 \text{ between } 2 \text{ and } 18\}$

Classify each set as empty or not empty.

23. {even prime numbers}
24. {integers between 2 and 3}
25. {real numbers between 2 and 3}
26. $\{x: x \in N, 3x + 2 = 9\}$

Insert the correct symbol $\subseteq$, $\subset$, $\nsubseteq$ or $\not\subset$ between the following pairs of sets.

27. $\{-2\}$ {integers}

28. $\{3, 5\}$ {x: x is an even number}

29. {x: x is an even number between -1 and 3} $\{-2, -1, 0, 1, 2, 4, 5\}$

30. {x: x is an integer} {x: x is a rational number}

31. {x: x is a vowel} {x: x is a consonants}

32. {x: x is an integer greater than or equal to zero} {x: x is a whole number}

33. {x: x is an irrational number} {x: x is a rational number}

List all subsets of each set.

34. $\{2\}$
35. $\{0, 1\}$
36. $\{-2, 1, 3\}$

Find the number of subsets of each set.

37. {x: x is a multiple of 2 between 7 and 15}

38. {x: x is a prime number between 1 and 12}

39. $\{-3, -2, -1, 0, 1, 2, 3\}$

40. {x: x is an odd number between 3 and 18}

1.2 Operation of Sets

Union of Sets

The union of two sets A and B, denoted by A ∪ B, is the set having all elements which belong to either A or B or both A and B.

Example

Given that A = {1, 2, 4, 5, 7} and B = {1, 3, 4, 6, 7, 8}, find A ∪ B.

A ∪ B = {1, 2, 3, 4, 5, 6, 7, 8}

Intersection of Sets

The intersection of two sets A and B, denoted by *A* ∩ *B*, is the set having all elements which belong to both A and B.

Example

Given that A = {1, 2, 3, 6, 7, 8, 10} and B = {2, 3, 5, 8, 9, 10, 12}, find *A* ∩ *B*.

A ∩ B = {2, 3, 8, 10}

If there are no common elements between two sets A and B, then A ∩ B = ∅. Such sets are called disjoint sets.

The Complement of a Set

The complement of a set A is the set having all elements that belong to the universal set but not to A.

Example

Given that U = {1, 2, 3, 4, 5, 6, 7, 8, 9, 10} and A = {1, 3, 5, 7, 9}, find A'.

A' = {2, 4, 6, 8, 10}

Exercise 1.2

1. Let A = {3, 4, 7, 9} and B = {1, 2, 3, 5, 6, 9, 10}. Find:

    (a) A ∪ B   (b) B ∪ A

2. Let A = {x: x is an integer between −5 and 3} and B = {x: x is an integer between −3 and 6}. Find:

    (a) A ∪ B   (b) B ∪ A

3. Let A = {1, 2, 5, 10} and B = {0, 1, 7, 10}. Find:

(a) A ∪ B (b) B ∪ A

4. Let A = {− 2, − 1, 0, 1, 7, 10} and B = {− 1, 0, 1, 2, 7, 9}. Find:

   (a) A ∩ B (b) B ∩ A

5. Let A = {x: x is a whole number between − 3 and 8} and B = {x: x is a prime number between 1 and 15}. Find:

   (a) A ∩ B (b) B ∩ A

6. Let A = {− 2, − 1, 3, 5}, B = {− 3, − 1, 0, 4} and C = {− 1, 2, 6}. Find:

   (a) (A ∪ B) ∪ C (b) A ∪ (B ∪ C)

7. Let A = {1, 2, 3, 4, 5, 6}, B = {2, 3} and C = {3, 4, 5}. Find:

   (a) (A ∩ B) ∩ C (b) A ∩ (B ∩ C)

8. Let A = {−2, 0, 1, 3, 4}, B = {−1, 1, 2, 5} and C = {−2, 0, 2, 5}. Find:

   (a) A ∪ (B ∩ C) (b) (A ∪ B) ∩ (A ∪ C)

9. Let A = {−3, −1, 1, 2, 5}, B = {−2, 0, 1, 3} and C = {−4, −1, 2, 5}. Find:

   (a) A ∩ (B ∪ C) (b) (A ∩ B) ∪ (A ∩ C)

10. Let U = {1, 2, 3, 4, 5, 6, 7, 8, 9, 10}, A = {1, 3, 4, 6, 7, 9} and B = {2, 4, 5, 7, 10}. Find:

    (a) A′ (b) B′

11. Let U = {1, 2, 3, 4, 5, 6, 7, 8, 9, 10, 12}, A = {1, 3, 5, 12} and B = {2, 4, 5, 7, 10}. Find:

    (a) A′ (b) B′

12. Let U = {a, b, c, d, e}, A = {c, e} and B = {a, b, c}. Find:

    (a) (A ∪ B)′ (b) A′ ∩ B′

13. Let U = {1, 2, 3, 4, 5, 6, 7, 8, 9, 10}, A = {1, 3, 5, 6, 10} and B = {1, 2, 3, 4, 6, 10}. Find:

    (a) (A ∩ B)′ (b) A′ ∪ B′

14. Let ξ = {students in a school}, A = {students who study mathematics} and B = {girls in the science class}. Express the following statement in set notation.

"All girls in the science class study mathematics".

15. Let ξ = {people who play football}, P = {professionals who play football} and R = {rich people who play football}. Express the following statement in a set notation.

"Not all professionals who play football are rich".

16. Let ξ = {people who take a driving test}, A = {students who take a driving test} and B = {students who failed a driving test}. Express the following statement in a set notation.

"None of the student failed a driving test".

17. Let ξ = {students in a school}, H = {students who study history}, P = {students who study physics}. Express the following statement in a set notation.

"There are no students who study both history and physics".

18. Let ξ = {children who are members of a club}, G = {girls who are members of the club) and R = {children in the club who like reading}. Express the following statement in a set notation.

"Most girls in the club like reading".

1.3 Venn Diagrams

Relationship between sets can be express visually by drawing diagrams called Venn Diagrams. A Venn diagram consists of a closed shape, usually circles, enclosed in a rectangle. The rectangle represents the universal set and the closed shapes represent subsets of the universal set.

Classification

Two Sets

8 CHAPTER 1 Sets

Two overlapping sets divide the universal set into four regions as shown in Figure 1.1.

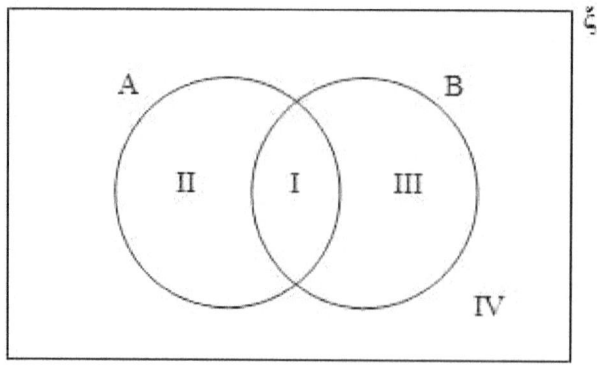

Figure 1.1

The various regions are named in terms of set A and set B as follows.

I: A ∩ B   II: A ∩ B'   III: A' ∩ B   IV: (A ∪ B)' or A' ∩ B'

Three Sets

Three overlapping sets A, B and C divide the universal set into eight regions as indicated in Figure 1.2.

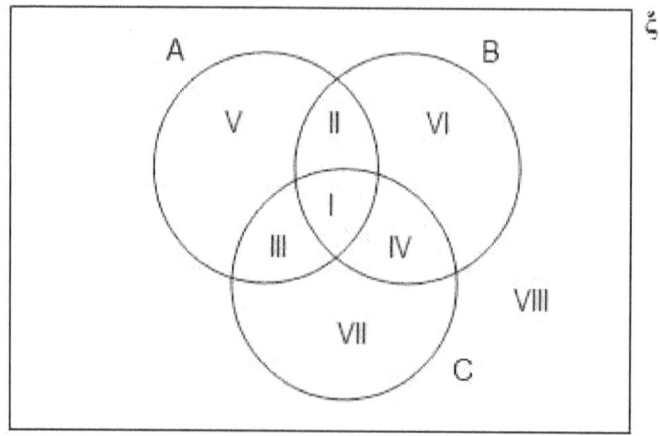

Figure 1.2

The names of the regions in terms of sets A, B and C are as follows.

I: A ∩ B ∩ C   II: A ∩ B ∩ C'   III: A ∩ B' ∩ C

IV: A' ∩ B ∩ C   V: A ∩ B' ∩ C'   VI: A' ∩ B ∩ C'

VII: A' ∩ B' ∩ C   VIII: (A ∪ B ∪ C)' or A' ∩ B' ∩ C'

Example

Draw a Venn diagram representing three overlapping sets P, Q and R. Shade the region corresponding to the set P ∪ (Q ∩ R).

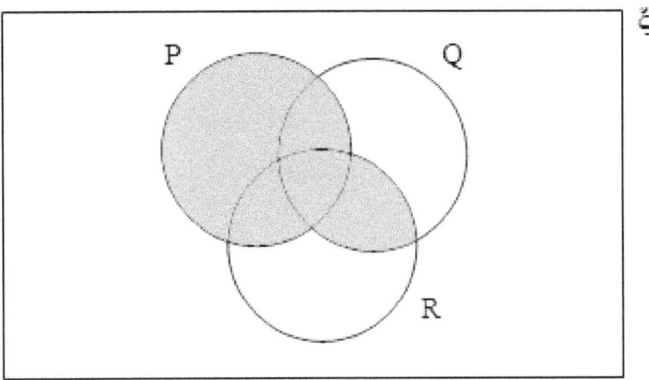

Exercise 1.3

1. Draw a Venn diagram with two overlapping sets A and B. Shade the region corresponding to the following sets.

   (a) A ∩ B   (b) A ∪ B   (c) A' ∩ B'

2. Draw a Venn diagram with two overlapping sets A and B. Shade the region corresponding to the following sets.

   (a) A' ∪ B   (b) A ∩ B'   (c) A' ∪ B'

3. Draw a Venn diagram with three overlapping sets A, B and C. Shade the region corresponding to the following sets.

   (a) A ∩ (B ∪ C)   (b) A ∩ (B' ∩ C)'

4. Draw a Venn diagram with three overlapping sets, A, B and C. Shade the region corresponding to the following sets.

   (a) A ∩ (B ∪ C)'   (b) (A' ∪ B) ∩ C

CHAPTER 1 Sets

Review Exercise 1

List the elements of the following sets between curly brackets.

1. The set of positive integers through 6 and 12.

2. The set of multiples of 3 between 8 and 15.

3. The set of prime numbers between 0 and 3.

4. $\{x: x \text{ is a factor of } 36\}$

5. $\{x: x \text{ is a square of integers between } -4 \text{ and } 5\}$

Classify each statement as true or false.

6. $5 \in \{-2, -1, 2, 5, 7\}$

7. $8 \notin \{x: x \text{ is a prime number}\}$

8. $4 \in \{x: x \text{ is a factor of } 6\}$

9. $2 \notin \{x: x \text{ is an irrational number}\}$

10. $-9 \in \{x: x \text{ is a whole number}\}$

11. $7 \in \{x:x \text{ is a factor of } 21\}$

Classify each statement as finite or infinite.

12. $\{x: x \text{ is a factor of } 18\}$

13. $\{x: x \text{ is a real number between } 5 \text{ and } 8\}$

14. $\{x: x \text{ is a multiple of } 3 \text{ between } 10 \text{ and } 50\}$

15. $\{x: x \text{ is an even number}\}$

Find the number of elements of each set.

16. $\{x: x \text{ is a factor of } 36\}$

17. $\{x: x \text{ is an integer from } -3 \text{ to } 1\}$

18. $\{x: x \text{ is a multiple of } 5 \text{ between } 12 \text{ and } 28\}$

19. {x: x is a vowel}

Classify each set as empty or not empty.

20. {x: x is an irrational number}

21. {Whole numbers between −2 and 0}

22. {x: x is a prime number between 1 and 3}

23. {x: x ∈ R, 3 < x < 4}

Insert the correct symbol ∈, ∉, ⊆, ⊂ or ⊄ between the following pairs of sets.

24. 4   {1, 2, 3, 4, 5}

25. 6   {2, 3, 5}

26. {2, 3, 5}   {1, 2, 4}

27. {2, 3, 5}   {1, 2, 3, 4, 5}

28. {x: x is a whole number}   {x: x is a non-negative integer}

29. 30   {x: x is a multiple of 5}

30. 9   {x: x is a prime number}

31. {x: x is an odd number between 15 and 25}   {x: x is a whole number}

32. Let A = {3, 4, 6, 8}, B = {3, 5, 7, 8} and C = {1, 2, 4, 5, 7}. Find:

   (a) A ∪ B   (b) B ∪ C   (c) A ∪ C

33. Let A = {−2, −1, 0, 1, 2, 3}, B = {−3, −1, 0, 3, 5} and C = {−4, −2, −1, 3}. Find:

   (a) A ∩ B   (b) A ∩ C   (c) B ∩ C

34. Let U = {1, 2, 3, 4, 5, 6, 7, 8, 9, 10}, A = {2, 3, 4, 5, 6}, B = {5, 6, 7, 8, 10}.

   (a) A′   (b) B′

   (c) (A ∪ B)′   (d) (A ∩ B)′

35. Draw a Venn diagram with three overlapping sets, A, B and C. Shade the region corresponding to the following sets.

(a) $A' \cap B \cap C'$

(b) $(A \cup B) \cap C'$

(c) $(A \cap B) \cap C'$

2 Functions

2.1 Defining a Function

A function from a set A to a set B, written briefly as $f : A \to B$ is a relation that assigns to each element in set A, called the domain, exactly one element in the set B, called the co-domain (or the range).

The mapping diagram, shown in Figure 2.1, represents a function f.

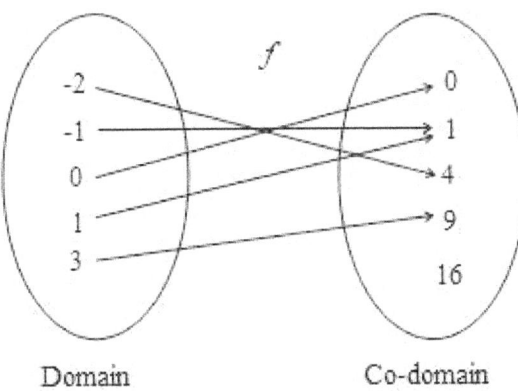

Figure 2.1

For any element x in the domain the corresponding element y in the co-domain is called the image, denoted by $f(x)$, read "f of x". You can see from Figure 2.1 that $f(-2) = 4, f(-1) = 1, f(0) = 0, f(1) = 1$ and $f(3) = 9$. The set of all images is called the range. In this case the range is $\{0, 1, 4, 9\}$. The elements of the range of a function are also called values of a function.

A function can be given in different representation such as a list, a table, a graph, an equation or rule.

Properties of Functions

1. Each element in the domain is matched with an element in the co-domain.

2. Some element in the co-domain may not be matched with any element in the domain.

3. Two or more elements in the domain may be matched with the same element in the co-domain.

4. An element in the domain cannot be matched with two different elements in the co-domain.

Finding the Domain and Range of a Function

You can find the domain of a function by identifying all of the possible x-values of the function, and the range by identifying all of the possible y-values of the function.

Examples

Find the domain and range of each function

1. $f(x) = 2x + 3$

    We can substitute any number for x, so the domain is all real numbers.

    The function can take on any value and so the range is all real numbers.

2. $f(x) = x^2 - 3$

    We can substitute any number for x, so the domain is all real numbers.

    The graph of this function is a parabola that opens upwards, and its vertex is at $(0, -3)$. The minimum value of this function is -3. Because the graph opens upwards and extends infinitely from -3, the range is all real numbers greater than or equal to -3.

3. $f(x) = \sqrt{x + 5}$

    We cannot take the square root of a negative number. This means x can take on numbers -5 or more. So the domain is all real numbers greater than or equal to -5.

    The square root of a number is either zero or positive. So, the range is all non-negative real numbers.

You can determine whether a graph represents a function by using the vertical line test.

Vertical Line Test

A graph represents a function if no vertical line intersects the graph of the function at more than one point.

Function Notation

If a function is defined by an equation, you can name the function by a letter such as f, g or h. For example, the equation $y = 3x + 2$ that describes y as a function of x can be given the name "f" and written as $f(x) = 3x + 2$ or $f : x \rightarrow 3x + 2$.

Evaluating a Function

The process of finding the value of $f(x)$ for a given value of x is called evaluating a function. You can do this by substituting a given x- value into the equation to obtain the value of $f(x)$..

Examples

1. Evaluate the function $f(x) = 5x - 27$ at $x = 3$.

$$f(x) = 5x - 27$$
$$f(3) = 5(3) - 27$$
$$= 15 - 27$$
$$= -12$$

2. Evaluate the function $f(x) = -3x^2 + 19$ at $x = -2$.

$$f(x) = -3x^2 + 19$$
$$f(-2) = -3(-2)^2 + 19$$
$$= -12 + 19$$
$$= 7$$

16 CHAPTER 2 Functions

Exercise 2.1

1. Determine whether each relation is a function.

(a)

Domain → Range: {1,2,3} → {1,3,4}

(b)

Domain → Range: {1,2,0} → {-2,-1,0}

(c)

Domain → Range: {-1,0,1} → {2,3,4}

(d)

Domain → Range: {0,1,3} → {3}

2. Determine whether each graph is that of a function.

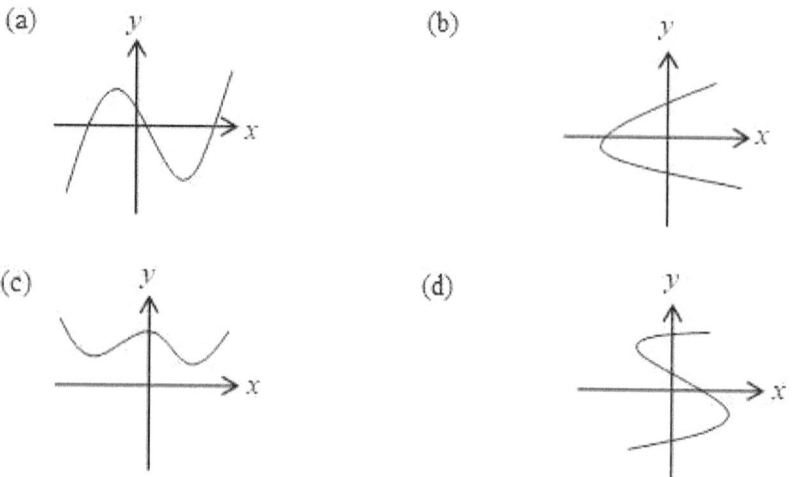

3. Find the domain and range of each of the following functions.

(a)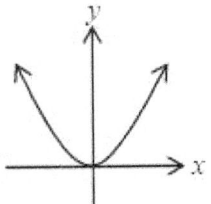

(b)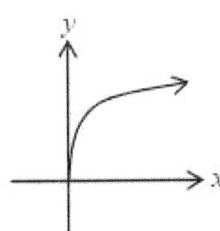

(c)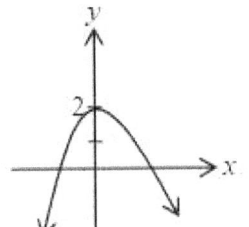

(d)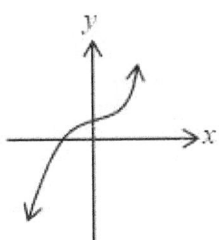

4. Find the domain and range of each function.

   (a) $f(x) = \sqrt{x} - 2$ (b) $g(x) = 2x^3 + 3$

   (c) $h(x) = 5 - 3x$ (d) $f(x) = |x - 1| + 3$

5. Let $f(x) = 3x^2 - 2x + 5$. Find each value of the function.

   (a) $f(-1)$ (b) $f(0)$

   (c) $f(2)$ (d) $f(3)$

6. Let $g(x) = 3 + 2x - x^2$. Find each value of the function.

   (a) $g(-2)$ (b) $g(-1)$

   (c) $g(1)$ (d) $g(2)$

7. Let $f(x) = (3x + 2)/(x - 5)$, where $x \neq 5$. Find each value of the function.

   (a) $h(-3)$ (b) $h(1)$

   (c) $h(6)$ (d) $h(8)$

18 CHAPTER 2 Functions

8. Let $f(x) = \sqrt{3x^2 - 2}$. Find each value of the function.

   (a) $f(-3)$ (b) $f(-1)$

   (c) $f(3)$ (d) $f(11)$

9. The function $g(x)$ is defined by $g: x \to 2 - 3x$. Solve the equation $g(x) = 8$.

10. The function $h(x)$ is defined by $h(x) = x^2$. Solve the equation $h(a + 1) = 16$.

11. The function $f(x) = (3x + 2)/(x - 3)$, where $x \neq 3$, is defined on the set of real numbers. Solve the equation $f(x) = 5/4$.

12. The function $g(x)$ is defined by $g(x) = \sqrt{x - 2}$, where $x \geq 2$. Solve the equation $g(x) = 3$.

2.2 Composition of Functions

Two functions f and g can be combined to form another function called the composition of f with g, denoted by $f \circ g$. The composition of the functions f and g is given by

$$(f \circ g)(x) = f(g(x)).$$

$(f \circ g)$ is called the composite function.

Examples

1. Let $f(x) = 3x + 4$ and $g(x) = 2x - 3$. Find the composition of f with g.

$$(f \circ g)(x) = f(g(x)) \quad \text{Definition of } f \circ g$$
$$= f(2x - 3) \quad \text{Replace } g(x) \text{ with } 2x - 3$$
$$= 3(2x - 3) + 4 \quad \text{Substitute } 2x - 3 \text{ for } x$$
$$= 6x - 5 \quad \text{Simplify}$$

2. Let $f(x) = -2x + 3$ and $g(x) = 5 - 4x$. Find the value of $(g \circ f)(2)$.

$(gof)(x) = g(f(x))$ Definition of $g \circ f$

    $= g(-2x + 3)$ Replace $f(x)$ with $-2x + 3$

    $= 5 - 4(-2x + 3)$ Substitute $-2x + 3$ for x

    $= 8x - 7$ Simplify

$(gof)(2) = 8(2) - 7 = 9$ Substitute 2 for x

You can obtain the same result as follows.

$(gof)(2) = g(f(2))$ Definition of $g \circ f$

    $= g(-2(2) + 3)$ Substitute 2 for x in $-2x + 3$

    $= g(-1)$ Simplify

    $= 5 - 4(-1)$ Substitute -1 for x in $5 - 4x$

    $= 9$ Simplify

3. Let $f(x) = 2x + 3$. Find the composition $(f \circ f)(x)$.

$(fof)(x) = f(f(x))$

    $= f(2x + 3)$

    $= 2(2x + 3) + 3$

    $= 4x + 9$

Note that $f(f(x))$ can be written as $f^2(x)$.

The composition of f and g is generally not the same as the composition of g with f.

Exercise 2.2

1. Functions f and g are defined, for $x \in R$, by $f(x) = 3x + 2$ and $g(x) = x^2 - 1$. Find

  (a) $f(g(x))$ (b) $g(f(x))$

2. Functions f, g and h are defined, for $x \in R$, by $f(x) = 2x + 3$, $g(x) = 2x^2$ and $h(x) = 2 - 3x$. Find

   (a) $fg(x)$ (b) $fh(x)$

   (c) $gh(x)$ (d) $gf(x)$

3. Functions f, g and h are defined, for $x \in R$, by $f(x) = 3 - x^2$, $g(x) = 2x + 5$ and $h(x) = x^2 + 1$. Find

   (a) $hf(x)$ (b) $gf(x)$

   (c) $hg(x)$ (d) $f^2(x)$

   (e) $g^2(x)$ (f) $h^2(x)$

4. Functions $f(x) = x^2$, $g(x) = 3x + 2$ and $h(x) = x - 2$ are defined on the set of real numbers, R. Find:

   (a) $fgh(x)$ (b) $ghf(x)$

   (c) $hfg(x)$ (d) $hgf(x)$

5. The functions $f(x) = 3x - 1$ and $g(x) = x^2$ are defined on the set of real numbers, R. Find:

   (a) $fg(3)$ (b) $gf(5)$ (c) $fg(-2)$

6. Functions f and g are defined, for $x \in R$, by $f(x) = (2x - 1)/5$ and $g(x) = (3x + 2)/(x - 1)$, where $x \neq 1$. Find:

   (a) $gf(2)$ (b) $fg(2)$

7. The functions f and g are defined on the set of real numbers, R, by $f(x) = 2x + 3$ and $g(x) = x^2 - 1$. Find:

   (a) $fg(-2)$ (b) $gf(2)$

   (c) $f^2(-1)$ (d) $g^2(3)$

8. The functions f and g are defined, for $x \in R$, by $f(x) = 2x + 1$ and $g(x) = 3x - 2$. Solve the following equations.

   (a) $fg(x) = 9$ (b) $gf(x) = 19$

9. The functions f and g are defined, for $x \in R$, by $f(x) = x^2$ and $g(x) = 2x + 1$. Solve the following equations

   (a) $fg(x) = 9$ (b) $gf(x) = 51$

10. The functions f and g are defined, for $x \in R$, by $f(x) = 2x - 1$ and $g(x) = (3x + 2)/(x + 2)$, where $x \neq -2$. Solve the equation $fg(x) = 3x$.

2.3 One – to – one Functions

A function f is one-to-one if each element y in the range corresponds to exactly one element x in the domain.

We illustrate this in Figure 2.2.

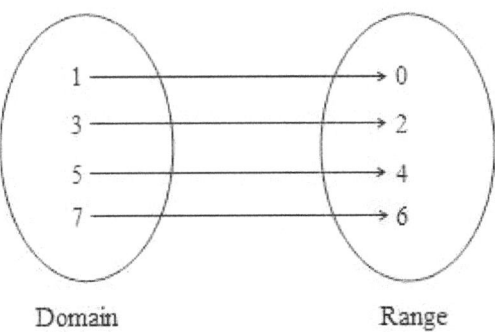

Domain   Range

Figure 2.2

Horizontal Line Test

A function is one-to-one if and only if every horizontal line intersects the graph of the function in at most one point.

Example

Determine whether if each function is one-to-one.

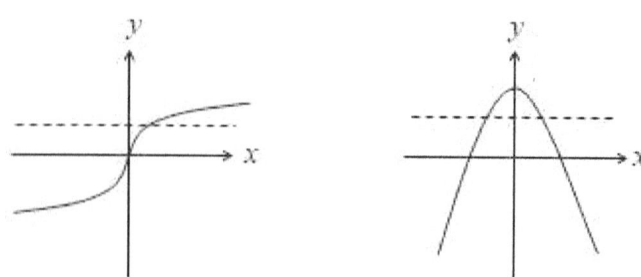

(a) Because no horizontal line intersects the graph of the function at more than one point, the function is one-to-one.

(b) Because it is possible to have a horizontal line that intersects the graph at more than one point, the function is not one-to-one.

The following is an equivalent definition of one-to-one function.

A function f is one-to-one if and only if $f(a) = f(b)$ implies $a = b$ for all a and b in the domain of f.

This definition is useful for proofs involving one-to-one functions.

Examples

1. The function $f(x) = 5x - 3$ is defined on the set of all real numbers, R. Show that the function is one-to-one.

$$\text{Let } a, b \in R.$$

$$f(a) = 5a - 3$$

$$f(b) = 5b - 3$$

Equating the two expressions and simplifying gives

$$5a - 3 = 5b - 3$$

$$a = b$$

Since $a = b$ when $f(a) = f(b)$, f is one-to-one.

2. The function g is defined on the set of real numbers, R, by $g(x) = 3x^2 + 2$. Determine whether or not the function $g(x)$ is one-to-one.

$$\text{Let } a, b \in R.$$

$$g(a) = 3a^2 + 2$$

$$g(b) = 3b^2 + 2$$

Equating the two expression and simplifying gives

$$3a^2 + 2 = 3b^2 + 2$$

$$a^2 = b^2$$

$$a = \pm b$$

Observe that a has two different values of b. Hence, the function g is not one- to –one.

Exercise 2.3

1. Determine whether each function is one-to-one.

(a)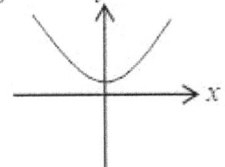

(b)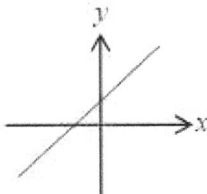

(c)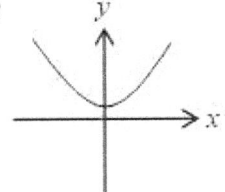

(d)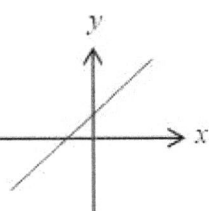

(e)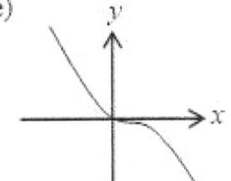

(f)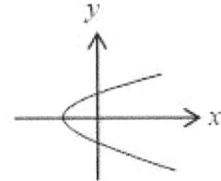

2. Determine whether each of the following functions defined on the set of real numbers, R, is one-to-one.

   (a) $f(x) = 7x + 5$

   (b) $f(x) = \frac{1}{4}x - 3$

   (c) $h(x) = \frac{3x}{2x-1}, x \neq \frac{1}{2}$

   (d) $g(x) = \frac{3x+2}{x+4}, x \neq -4$

3. Determine whether or not the following functions are one-to-one.

   (a) $f(x) = 3x^2 - 2$

   (b) $g(x) = 5x + 3$

   (c) $g(x) = \frac{5}{x+4}, x \neq -4$

   (d) $h(x) = x^3 - 1$

   (e) $f(x) = \frac{2x+3}{x-5}, x \neq 5$

   (h) $h(x) = \frac{2}{x^2-1}, x \neq \pm 1$

2.4 Inverse Functions

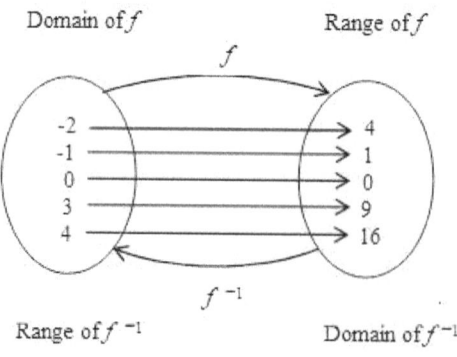

Figure 2.3

Figure 2.3 shows a function f from set A to set B. The backward mapping from set B to set A forms another function, called the inverse function of f, denoted by f^{-1}. Observe that the domain of f is the range of f^{-1} and the range of f is the domain of f^{-1}.

The function shown in Figure 2.4 is a many-to-one function.

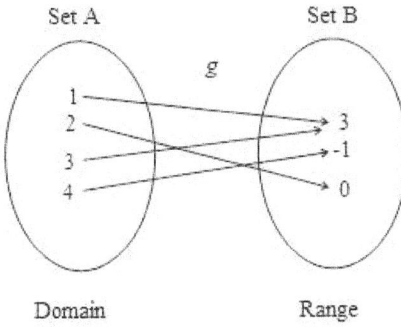

Figure 2.4

The backward mapping does not form an inverse function since an element in set B corresponds to two elements in set A.

A function f has an inverse f^{-1} if and only if f is one-to-one. It follows that, a function f has an inverse function f^{-1} if and only if no horizontal line intersects the graph of f at more than one point.

Definition of Inverse Functions

If f and g are two functions such that $f(g(x)) = x$ and $g(f(x)) = x$ then f and g are inverse function of each other.

Example

Show that $f(x) = 3x + 5$ and $g(x) = (x - 5)/3$ defined on the set of real numbers, R, are inverse functions of each other.

You need to show that $f(g(x)) = x$ and $g(f(x)) = x$.

$$f(g(x)) = f\left(\frac{x-5}{3}\right)$$

$$= 3\left(\frac{x-5}{3}\right) + 5$$

$$= x - 5 + 5$$

$$= x$$

$$g(f(x)) = g(3x+5)$$
$$= \frac{3x+5-5}{3}$$
$$= \frac{3x}{3}$$
$$= x$$

The two functions are inverses of each function.

Finding an Inverse Function

The following examples illustrate the steps for finding an inverse function algebraically.

Examples

Determine the inverse function of each of the following functions, defined on the set of real numbers, R.

1. $f(x) = 4x + 3$

$f(x) = 4x + 3$	Origin function
$y = 4x + 3$	Replace $f(x)$ with y
$x = 4y + 3$	Interchange x and y
$y = \frac{x-3}{4}$	Solve for y
$f^{-1}(x) = \frac{x-3}{4}$	Replace y with $f^{-1}(x)$

2. $f(x) = \frac{2x-3}{x+2}$, where $x \neq -2$.

$y = \frac{2x-3}{x+2}$	Replace $f(x)$ with y
$x = \frac{2y-3}{y+2}$	Interchange x and y
$y = \frac{2x+3}{2-x}$	Solve for y
$f^{-1}(x) = \frac{2x+3}{2-x}$	Replace y with $f^{-1}(x)$

Graphs of Inverse Functions

Recall that, to obtain an inverse function we interchange x and coordinates in the ordered pair (x, y). If the point (a, b) is on the graph of f, then the point (b, a) is on the graph of f^{-1}. This means that the graphs of a function f and its inverse function f^{-1} are reflection of each other in the line $y = x$.

You can sketch the graph of an inverse function without knowing the equation of the inverse function. Simply find the coordinates of points that lie on the original function, then interchange the x- and y- coordinates and plot these points and sketch the graph of the inverse function.

Example

Sketch the graphs of the function $f(x) = 2x + 6$ and its inverse function on the same axis.

The graph of f has x-intersect at $(-3, 0)$ and y-intersect at $(0, 6)$. Plot these points and draw a line through them to obtain the graph of f. Next, interchange the coordinates of the points $(-3, 0)$ and $(0, 6)$. We get $(0, -3)$ and $(6, 0)$. Plot these points and draw a line through them to obtain the graph of the inverse function $f^{-1}(x)$..

A graph of the function f and its inverse function f^{-1} is shown in Figure 2.5.

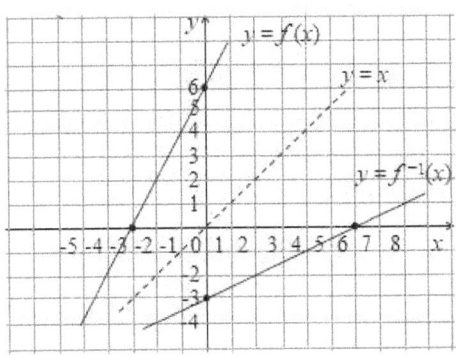

Figure 2.5

Exercise 2.4

1. Determine whether each of the following functions has an inverse.

(a)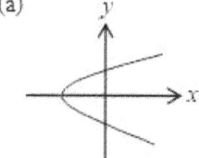

(b)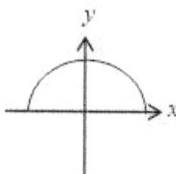

(c)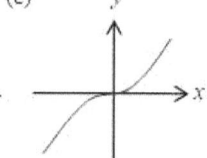

(d)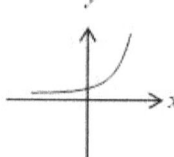

(e)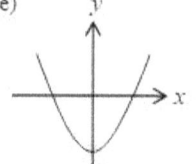

(f)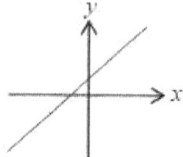

2. Determine whether the functions f and g are inverse functions of each other.

   (a) $f(x) = 3 - 2x, g(x) = \frac{1}{2}(3 - x)$

   (b) $f(x) = -\frac{1}{4}x + 3, \ g(x) = 12 - 4x$

   (c) $f(x) = 3x - 1, \ g(x) = \frac{1}{2}(x + 1)$

   (d) $f(x) = \frac{1}{x+2}, x \neq -2, \ g(x) = \frac{2-x}{x}, x \neq 0,$

   (e) $f(x) = 5 - 4x, g(x) = \frac{1}{4}(5 - x)$

3. Find the inverse function of each of the following functions.

   (a) $f(x) = 3x - 8$

   (b) $g(x) = 4 - x^2, x \geq 0$

   (c) $h(x) = \frac{5}{2x+1}, x \neq -\frac{1}{2},$

   (d) $f(x) = \frac{2x+3}{x-4}, x \neq 4$

   (e) $g(x) = \sqrt[3]{x+2}$

   (f) $h(x) = (x+3)^2, x \geq -3$

4. On the same axes, sketch the graphs of the function f and its inverse f^{-1}.

   (a) $f(x) = x + 3$ (b) $f(x) = x^2 + 2, x \geq 0$

   (c) $f(x) = (x - 3)^2, x \geq 3$ (d) $f(x) = x^3 - 1$

   (e) $f(x) = |x - 5|, x \geq 5$ (f) $f(x) = \sqrt[3]{x} + 2$

2.5 Absolute Value Functions

A function that is written within an absolute value symbol is called absolute value function. Examples include $f(x) = |3x + 2|$ and $g(x) = |x^2 - 1|$.

Consider the function $f(x) = |x|$ defined as

$$f(x) = \begin{cases} -x \text{ if } x < 0 \\ x \text{ if } x \geq 0 \end{cases}$$

The graph of $f(x) = |x|$ is shown Figure 2.6.

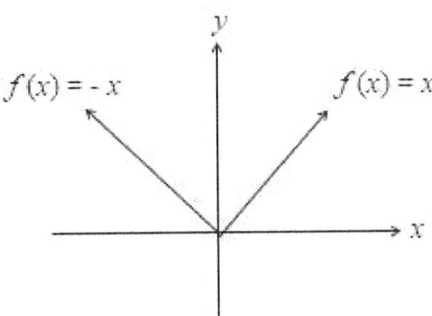

Figure 2.6

Observe the following.

1. The domain of the function is the set of all real numbers.

2. The range is the set of all real numbers greater than or equal to zero.

3. The vertex of the graph is (0, 0).

Absolute Value Equations

An absolute value equation is an equation that contains expressions in absolute value sign. Examples include $|2x + 1| = 9$ and $|5 - x| = 2|x - 3|$.

Solving Absolute Value Equations

The absolute value of a number is its distance from zero on a number line.

Consider the absolute value equation

$$|x| = 5$$

This equation has exactly two solutions: $x = -5$ and $x = 5$.

The following examples show how to solve absolute value equations.

Examples

Solve the following equations.

1. $|2x + 5| = 13$ Original equation

   The equation is equivalent to two linear equations $2x + 5 = -13$ and $2x + 5 = 13$. We solve these equations separately.

$2x + 5 = -13$	or	$2x + 5 = 13$
$2x = -18$		$2x = 8$
$x = -9$		$x = 4$

   You need to check your answers in the original equation. An absolute value equation may have no solution or may have only one solution.

   Check.

$|2x + 5| = 13$	$|2x + 5| = 13$
$|2(-9) + 5| = 13$	$|2(4) + 5| = 13$
$|-13| = 13$	$|13| = 13$

   The solutions are $x = -9$ and $x = 4$.

2. $|3x + 2| - 6 = x$

$\quad\quad |3x + 2| - 6 = x \quad\quad$ Original equation

First isolate the absolute value.

$\quad\quad |3x + 2| = x + 6$

Separate the equation into two linear equations and solve each equation separately.

$\quad\quad 3x + 2 = x + 6 \quad$ or $\quad 3x + 2 = -(x + 6)$

$\quad\quad\quad 2x = 4 \quad\quad\quad\quad\quad\quad 4x = -8$

$\quad\quad\quad x = 2 \quad\quad\quad\quad\quad\quad\ x = -2$

The solutions are $x = 2$ and $x = -2$. Check these solutions in the original equation.

3. $|x + 3| + 9 = 4$

$\quad\quad |x + 3| + 9 = 4 \quad\quad$ Original equation

First isolate the absolute value.

$\quad\quad |x + 3| = -5$

The absolute value of a real number cannot be negative. Thus, the equation has no solution.

4. $|2x + 1| = |x + 8|$

$\quad\quad |2x + 1| = |x + 8| \quad\quad$ Original equation

The equation is equivalent to

$\quad\quad 2x + 1 = x + 8 \quad$ or $\quad 2x + 1 = -(x + 8)$

$\quad\quad\quad x = 7 \quad\quad\quad\quad\quad\quad 2x + 1 = -x - 8$

$\quad\quad\quad\quad\quad\quad\quad\quad\quad\quad\quad\quad\quad x = -3$

The solutions are $x = -3$ and $x = 7$. Check these solutions in the original equation.

You can obtain the same results as follows.

$$|2x+1| = |x+8| \quad \text{Original equation}$$

Square both sides of the equation.

$$4x^2 + 4x + 1 = x^2 + 16x + 64$$

Simplifying this equation we get

$$x^2 - 4x - 21 = 0$$

Solving the quadratic equation gives

$$(x-7)(x+3) = 0$$

$$x = 7 \quad \text{or} \quad x = -3$$

The solutions are $x = -3$ or $x = 7$

Exercise 2.5

Solve the following equations.

1. $|x+2| = 6$

2. $|x+5| = 7$

3. $|4x-30| = 16$

4. $|5-7x| = 19$

5. $|2x+5| = -16$

6. $|22-7x| = 50$

7. $|3x-1| = -5$

8. $|6x-4| - 10 = -8$

9. $2|4-3x| - 6 = -2$

10. $|x+8| = |2x+1|$

11. $|2x-3| = |x-2|$

12. $|3x + 7| = |2x + 8|$

13. $|4x - 10| = 2|2x + 3|$

14. $3|3x - 2| = |9x - 12|$

15. $|3x + 10| = |x - 7|$

16. $|5x - 7| = 2|3x - 2|$

17. $|32 - 3x| = |45 - 4x|$

18. $2|3x + 2| = |2x - 7|$

19. $|3x + 25| = |5x + 4|$

20. $|9x + 21| = 3|2 - 3x|$

Review Exercise 2

1. Determine whether each graph is that of a function.

(a)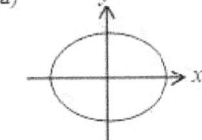

(b)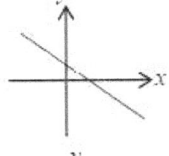

(c)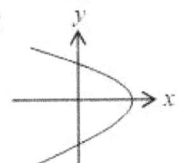

(d)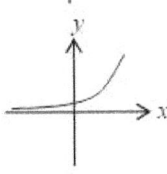

(e)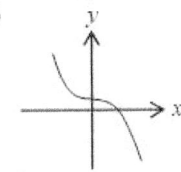

(f)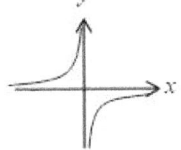

34 CHAPTER 2 Functions

2. Find the domain and range of each of the following functions.

(a)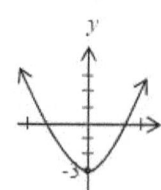

(b)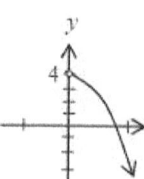

(c)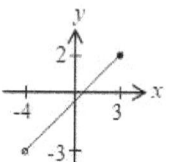

(d)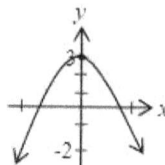

(e)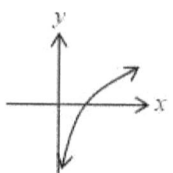

(f)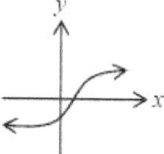

3. Find the domain and range of each function.

   (a) $f(x) = 2 - 3x$

   (b) $f(x) = |x - 5| - 2$

   (c) $f(x) = \sqrt{9 - 3x}$

   (d) $f(x) = 3x^2 + 2$

4. Given that $f(x) = 2x^2 + 3x - 7$, find the following values.

   (a) $f(-3)$ (b) $f(0)$ (c) $f(2)$

5. Given that $g(x) = 8 - 2x^3$, find the following values.

   (a) $g(-2)$ (b) $g(-½)$ (c) $g(2)$

6. Functions f and g are defined, for $x \in R$, by $f(x) = 2x - 3$ and $g(x) = 3x^2 + 2$, find

   (a) $fg(x)$ (b) $gf(x)$ (c) $f^2(x)$ (d) $g^2(x)$

7. The functions $f(x) = x^2 - 3x$, $g(x) = 2 - x^2$ and $h(x) = 3x + 2$ are defined on the set of real numbers, R.

(a) $fg(x)$ (b) $hf(x)$ (c) $gh(x)$

(d) $gf(x)$ (e) $fh(x)$ (f) $hg(x)$

(g) $f^2(x)$ (h) $g^2(x)$ (i) $h^2(x)$

8. The functions f and g are defined, for $x \in R$, by $f(x) = 5 - 4x$ and $g(x) = 3x^2$. Find

(a) $fg(-2)$ (b) $gf(0)$ (c) $gf(2)$ (d) $fg(3)$

9. Determine whether each function is one-to-one and so has an inverse function.

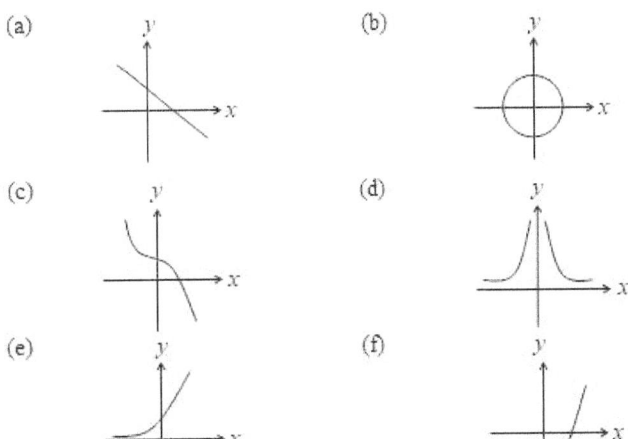

10. Determine whether the functions f and g are inverse functions of each other.

(a) $f(x) = 5x - 4$, $g(x) = \frac{1}{5}(x + 4)$

(b) $f(x) = \frac{1}{3}(x - 4)$, $g(x) = 3x + 4$

(c) $f(x) = \sqrt{x + 2}$, $g(x) = x^2 + 2, x \geq 0$

(d) (d) $f(x) = \sqrt[3]{x - 2}$, $g(x) = x^3 + 2$

11. Find the inverse function of each of the following functions.

(a) $g(x) = 7 + 3x$ (b) $h(x) = 2x^3 - 5$

(c) $f(x) = \frac{x}{x+3}$, $x \neq -3$ (d) $f(x) = \sqrt{x-3}$, $x \geq 3$

12. On the same axes, sketch the graphs of the function f and the inverse f^{-1}.

   (a) $f(x) = 3x + 2$ (b) $f(x) = 5 - 2x$

   (c) $f(x) = x^2 - 3$, ≥ 0 (d) $f(x) = x^2 - 1$, $x \geq 0$

13. Functions f and g are defined, for $x \in R$, by $f(x) = x^2 - 25$, $g(x) = 2x/(3x - 2)$

   (a) State the largest possible domain of each of the two functions.

   (b) Find the range of f.

   (c) Determine whether or not g is a one-to-one function.

14. A function g is defined, for $x \in R$, by $g(x) = \sqrt[3]{x+3}$. Find

   (a) g^{-1} (b) the range of g^{-1}

15. Functions f and g are defined, for $x \in R$, by $f(x) = 4/(x - 3)$, where $x \neq 3$, $g(x) = 2x + 1$. Find

   (a) $fg(x)$ (b) $gf(x)$ (c) $fg(-2)$ (d) $gf(1)$

16. Functions f and g are defined, for $x \in R$, by $f(x) = 2x^2 + 1$, $h(x) = x - 2$. Find the value of a for which $fh(a) = hf(a)$.

17. The functions $g(x) = x^2 - 1$ and $h(x) = (x + 2)/x$, where $x \neq 0$, are defined on the set of real numbers R. Find:

   (a) h^{-1} (b) $gh^{-1}(-½)$

18. Functions g and h are defined, for $x \in R$, by $g(x) = x^2 + p$, $h(x) = px + q$, where p and q are positive real numbers. Find the value of p and of q for which $gh(x) = 9x^2 + 12x + 7$.

19. Functions f and g are defined, for $x \in R$, by $f(x) = -2x + 3$, $g(x) = px + q$, where p and q are constants. Given that $g^{-1}(4) = 2$ and $gf(3) = -7$, find the value of p and q.

3 Quadratic Functions

3.1 Quadratic Equations

A quadratic equation is an equation that can be written in the general form $ax^2 + bx + c = 0$, where a, b and c are real numbers with $a \neq 0$. Examples include $x^2 + 4x - 3 = 0$, $2x^2 - 3x = 0$ and $3x^2 = 5x - 7$.

Solving Quadratic Equations

Quadratic equations can be solved algebraically by using the following methods: factorizing, completing the square and using the quadratic formula.

Solving Quadratic Equations by Factorizing

To solve a quadratic equation in the general form by factorizing, factorize the quadratic expression into two linear factors, set each factor to zero and then solve the resulting linear equations, as shown in the following examples.

Examples

Solve each of the following equations.

1. $x^2 - 5x - 24 = 0$

$x^2 - 5x - 24 = 0$	Original equation
$(x - 8)(x + 3) = 0$	Factorize left side of the equation
$x - 8 = 0$ or $x + 3 = 0$	Set each factor to 0
$x = 8 \qquad x = -3$	Solve each equation for x

2. $2x^2 = 12 - 5x$

   $\qquad 2x^2 = 12 - 5x \qquad$ Original equation

   Rewrite the equation in the general form.

$2x^2 + 5x - 12 = 0$	
$(2x - 3)(x + 4) = 0$	Factorize left side of the equation
$2x - 3 = 0$ or $x + 4 = 0$	Set each factor to 0
$x = 3/2 \qquad x = -4$	Solve each equation

CHAPTER 3 Quadratic Functions

Exercise 3.1(a)

Solve each equation.

1. $x^2 + 7x + 10 = 0$
2. $x^2 - 7x + 12 = 0$
3. $x^2 - 3x - 40 = 0$
4. $x^2 + 3x - 10 = 0$
5. $x^2 + 8x + 15 = 0$
6. $x^2 - 3x - 18 = 0$
7. $x^2 - 12x + 32 = 0$
8. $x^2 + 4x - 21 = 0$
9. $x^2 - 7x + 10 = 0$
10. $8 - 2x - x^2 = 0$
11. $42 + x - x^2 = 0$
12. $x^2 - 6x + 9 = 0$
13. $x^2 - 4x = 12$
14. $x^2 + 11x = -30$
15. $3x^2 - 4x - 20 = 0$
16. $2x^2 - 5x + 3 = 0$
17. $6x^2 = x + 1$
18. $4x^2 - 11x = 3$
19. $5x - 20x^2 = 0$
20. $3x^2 - 75 = 0$
21. $(x + 2)^2 = 16$
22. $(3x - 2)^2 = 25$
23. $(2x + 3)^2 = 16x^2$
24. $(x - 1)(x - 2) = 42$
25. $(x - 5)(x + 2) = 18$
26. $(x - 3)(x + 2) = 24$
27. $x(x - 10) = -24$
28. $8x(x + 1) = 30$
29. $(2x + 1)^2 = 3x^2 + 13$
30. $(3x - 2)^2 = x(6x - 4)$

Completing the Square

The method of completing the square is based on the relationship between the middle term and the constant term of the perfect square trinomial:

$$x^2 + 2ax + a^2 = (x + a)^2$$

$$x^2 - 2ax + a^2 = (x - a)^2$$

Notice that in each case the constant term is the square of one-half of the coefficient of the middle term. This relationship is true if and only if the coefficient of the square term is 1.

Solving Quadratic Equations by Completing the Square

When solving the quadratic equation $x^2 + bx + c = 0$, by completing the square you must add $(b/2)^2$ to both side of the equation in order to maintain equality. If the coefficient of the square is not 1, you must divide each side of the equation by the coefficient of the square term before you complete the square.

Examples

Solve each equation.

1. $x^2 + 6x - 40 = 0$

$$x^2 + 6x - 40 = 0 \qquad \text{Original equation}$$

$$x^2 + 6x = 40 \qquad \text{Add 40 to each side}$$

$$x^2 + 6x + 9 = 40 + 9 \qquad \text{Add } 3^2 = 9 \text{ to each side}$$

$$(x + 3)^2 = 49 \qquad \text{Factorize the left side}$$

$$x + 3 = \pm 7 \qquad \text{Take square root each side}$$

$$x = -3 \pm 7 \qquad \text{Subtract 3 from each side}$$

$$x = -3 + 7 \text{ or } x = -3 - 7$$

$$x = 4 \qquad x = -10$$

2. $2x^2 - 3x - 9 = 0$

$$2x^2 - 3x - 9 = 0 \qquad \text{Original equation}$$

$$2x^2 - 3x = 9 \qquad \text{Add 9 to each side}$$

$$x^2 - \frac{3}{2}x = \frac{9}{2} \qquad \text{Divide both sides by 2}$$

$$x^2 - \frac{3}{2}x + \frac{9}{16} = \frac{9}{2} + \frac{9}{16} \qquad \text{Add } \left(-\frac{3}{4}\right)^2 = \frac{9}{16} \text{ to each side}$$

$$\left(x - \frac{3}{2}\right)^2 = \frac{81}{16} \qquad \text{Factorize the left side}$$

$$x - \frac{3}{4} = \pm\frac{9}{4} \qquad \text{Take square root of each side}$$

$$x = \frac{3}{4} + \frac{9}{4} = 3 \quad \text{or} \quad x = \frac{3}{4} - \frac{9}{4} = -\frac{3}{2}$$

Exercise 3.1(b)

1. Find the constant that must be added to each binomial expression to make it a perfect square.

    (a) $x^2 - 12x$ (b) $y^2 - 14y$ (c) $x^2 + 8x$

    (d) $x^2 - 16x$ (e) $y^2 + 3y$ (f) $x^2 + 5x$

    (g) $x^2 - x$ (h) $x^2 - \frac{1}{2}x$ (i) $y^2 + \frac{1}{3}y$

2. Solve each equation by completing the square.

    (a) $x^2 + 12x - 2 = 0$ (b) $x^2 - 4x - 3 = 0$

    (c) $x^2 + 10x + 13 = 0$ (d) $x^2 + 3x - 17 = 0$

    (e) $x^2 + 3x - 27 = 0$ (f) $x^2 - 7x + 3 = 0$

    (g) $2x^2 + x - 2 = 0$ (h) $3x^2 - x = 6$

    (i) $3x^2 - 3x - 1 = 0$ (j) $4x^2 + 8x - 1 = 0$

    (k) $3x^2 - 2x - 12 = 0$ (l) $7x^2 - 2x - 3 = 0$

The Quadratic Formula

The solution of the quadratic equation $ax^2 + bx + c = 0$, where $a \neq 0$, are given by

$$x = \frac{-b \pm \sqrt{b^2 - 4ac}}{2a}$$

This solution is called the Quadratic Formula.

Quadratic Equations

Solving Equations by the Quadratic Formula

When using the Quadratic formula, first write the quadratic equation in general form in order to determine the values of a, b, and c.

Examples

Solve the following equations.

1. $x^2 + 4x - 45 = 0$

$$x^2 + 4x - 45 = 0 \quad \text{Original equation}$$

Identify the values of a, b and c in the quadratic equation.

Here, $a = 1$, $b = 4$ and $c = -45$.

By substituting these values into the quadratic formula gives

$$x = \frac{-4 \pm \sqrt{4^2 - 4(1)(-45)}}{2(1)}$$

$$x = \frac{-4 \pm \sqrt{196}}{2}$$

$$x = \frac{-4 \pm 14}{2}$$

$$x = 5 \text{ or } x = -9$$

2. $3x^2 = 7x - 2$

$$3x^2 = 7x - 2$$

Begin by writing the equation in the general form.

$$3x^2 - 7x + 2 = 0$$

Here $a = 3$, $b = -7$ and $c = 2$.

$$x = \frac{-(-7) \pm \sqrt{(-7)^2 - 4(3)(2)}}{2(3)}$$

$$x = \frac{7 \pm \sqrt{25}}{6}$$

$$x = \frac{7 \pm 5}{6}$$

$$x = 2 \text{ or } x = \frac{1}{3}$$

Exercise 3.1(c)

Solve the following equations.

1. $x^2 + 8x - 65 = 0$
2. $x^2 - 5x - 14 = 0$
3. $x^2 - 2x - 5 = 0$
4. $x^2 - 7x + 3 = 0$
5. $x^2 + 7x - 30 = 0$
6. $5x^2 + 4x - 1 = 0$
7. $16x^2 - 24x + 9 = 0$
8. $2x^2 - 6x + 1 = 0$
9. $3x^2 + 2x - 1 = 0$
10. $2x^2 - 3x - 1 = 0$
11. $2x^2 - \frac{1}{2}x - 5 = 0$
12. $3x^2 + \frac{1}{3}x - 3 = 0$
13. $3x^2 + 2x - \frac{3}{4} = 0$
14. $(x - 2)(x + 3) = 4$
15. $(x + 1)(2x - 4) = 7$
16. $3x - 5 = \frac{1}{x}$
17. $4x - \frac{1}{x} = 6$
18. $\frac{5}{x^2} + \frac{2}{x} = 1$
19. $2x - \frac{3}{x} - 3 = 0$
20. $\frac{6}{x^2} - \frac{2}{x} = 1$
21. $3 + \frac{8}{x} - \frac{5}{x^2} = 0$
22. $2x(x - 3) = x + 5$

3.2 Discriminant

The expression $b^2 - 4ac$ in the quadratic formula is called discriminant. The discriminant gives useful information about the nature of the roots of the quadratic equation $ax^2 + bx + c = 0$.

1. If $b^2 - 4ac > 0$, there are two distinct real roots.

2. If $b^2 - 4ac = 0$, there is one real root.

3. If $b^2 - 4ac < 0$, there are no real roots.

These conditions are illustrated in Figure 3.1.

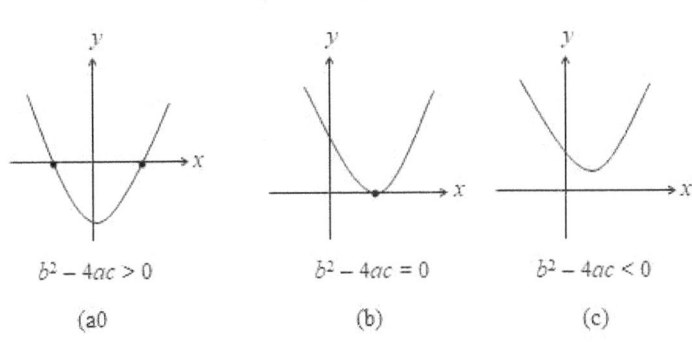

Figure 3.1

In case (a), the x-axis intersects the graph at two points, in case (b), the x-axis is a tangent to the curve and in case (c) the x-axis does not intersect the graph.

Examples

1. State the nature of the roots of the following quadratic equation.

    (a) $x^2 + 2x - 3 = 0$

    Here $a = 1$, $b = 2$ and $c = -3$.

    $b^2 - 4ac = 2^2 - 4(1)(-3) = 16$

    The equation has two distinct real roots.

    (b) $9x^2 + 6x + 1 = 0$

    $b^2 - 4ac = 6^2 - 4(9)(1) = 0$

    The equation has one real root.

    (c) $2x^2 - 3x + 5 = 0$

    $b^2 - 4ac = (-3)^2 - 4(2)(5) = -31$

    The equation has no real roots.

CHAPTER 3 Quadratic Functions

2. Find the range of values of k for which $x^2 + 2x - k = 0$ has real roots.

   Since the equation has real roots $b^2 - 4ac \geq 0$

   $$2^2 - 4(1)(-k) \geq 0$$

   $$4 + 4k \geq 0$$

   $$k \geq -1$$

3. Find the range of values of k for which the expression $2x^2 + 4x + k$ is always positive for all real values of x.

   The expression will always be positive when $b^2 - 4ac < 0$.

   $$4^2 - 4(2)(k) < 0$$

   $$16 - 8k < 0$$

   $$k > 2$$

Exercise 3.2

1. Find the nature of the roots of each of the following equations.

   (a) $x^2 - 4x + 3 = 0$
   (b) $x^2 - 3x + 5 = 0$
   (c) $x^2 - 4x + 4 = 0$
   (d) $2x^2 - 3x + 4 = 0$
   (e) $3x^2 + 7x - 6 = 0$
   (f) $4x^2 + 4x + 1 = 0$
   (g) $2x^2 - x + 5 = 0$
   (h) $3x^2 - x - 2 = 0$

2. Find the values of m for which the equation $x^2 + (m + 3)x + 4m = 0$ has equal roots.

3. Find the value of m if the equation $(2m + 1)x^2 + 3mx + m = 0$ has equal roots.

4. If the equation $px^2 + (p + 1)x + p = 0$ has equal roots, find the value(s) of p.

5. Find the range of values of k for which $kx^2 + 4x + 2 = 0$ has real roots.

6. Find the range of values of k for which $x^2 + x - k = 0$ has two distinct real roots.

7. Find the smallest possible integral value of k such that $2x + 3x + k = 0$ will have no real roots.

8. Find the range of values of k for which the equation $x^2 + (3 - k)x + 1 = 0$ has real roots.

9. Find the greatest possible integral value of k for which the equation $2x^2 - 7x + k = 0$ has real roots.

10. Find the range of values of k for which the expression $4x^2 + 8x + k$ is always positive for all real values of x.

3.3 Finding the Maximum or Minimum Value of Quadratic Functions

The quadratic expression $ax^2 + bx + c$ can be rewritten in the form $y = a(x - h)^2 + k$, by completing the square. In this form, we can identify the vertex as (h, k). The y-coordinate of the vertex is either the maximum or minimum value of the function.

Examples

1. Find the minimum value of $2x^2 - 5x + 7$.

$$2x^2 - 5x + 7 = 2\left[x^2 - \frac{5}{2}x + \frac{7}{2}\right]$$

$$= 2\left[x^2 - \frac{5}{2}x + \frac{25}{16} - \frac{25}{16} + \frac{7}{2}\right]$$

$$= 2\left[\left(x - \frac{5}{4}\right)^2 + \frac{31}{16}\right]$$

$$= 2\left(x - \frac{5}{4}\right)^2 + \frac{31}{8}$$

Because $(x - 5/4)^2$ is a square it is always positive or zero. If we substitute a value for x other than 5/4, the value of $2x^2 - 5x + 7$ becomes larger. The minimum value will occur when the square term is zero. Hence, the minimum value is 31/8, and this occur when $x = 5/4$.

CHAPTER 3 Quadratic Functions

2. Find the maximum value of $5 + 2x - 3x^2$.

$$5 + 2x - 3x^2 = -3x^2 + 2x + 5$$

$$= -3\left[x^2 - \frac{2}{3}x - \frac{5}{3}\right]$$

$$= -3\left[x^2 - \frac{2}{3}x + \frac{1}{9} - \frac{1}{9} - \frac{5}{3}\right]$$

$$= -3\left[\left(x - \frac{1}{3}\right)^2 - \frac{16}{9}\right]$$

$$= -3\left(x - \frac{1}{3}\right)^2 + \frac{16}{3}$$

The value of $-3(x - 1/3)^2$ is always negative or zero. If we substitute a value for x other than $1/3$, the value of $5 + 2x - 3x^2$ becomes smaller. The maximum value will occur when the square term is zero. Hence, the maximum value is $16/3$, and this occur when $x = 1/3$.

Exercise 3.3

1. By completing the square, find the minimum value of the following expressions.

   (a) $x^2 + 3x - 2$ (b) $x^2 - 8x + 6$

   (c) $x^2 + 2x + 3$ (d) $2x^2 + 3x - 4$

   (e) $3x^2 - 6x + 4$ (f) $2x^2 + 6x + 5$

2. By completing the square, find the maximum value of the following expressions.

   (a) $2 + x - x^2$ (b) $5 - 2x - x^2$

   (c) $6 + 5x - x^2$ (d) $3 - 5x - 2x^2$

   (e) $2 + 4x - 3x^2$ (f) $1 - 6x - 3x^2$

3. By completing the square, find the minimum value or maximum value of the following expressions. What value of x gives the maximum or minimum value?

   (a) $x^2 - 6x + 10$ (b) $6 - x - x^2$

(c) $2 - 2x - x^2$ (d) $2x^2 + 8x + 9$

(e) $3x^2 + 6x + 7$ (f) $-7 + 12x - 3x^2$

3.4 Sketching Graphs of Quadratic Functions

The graph of the quadratic function $y = ax^2 + bx + c$, where a, b and c are real numbers and $a \neq 0$, is a parabola.

Every parabola has an axis of symmetry. The axis of symmetry is a vertical line midway between any pair of symmetric points on the parabola, and intersects the parabola at its vertex. The vertex is either the minimum or maximum point on a parabola.

If $a > 0$, the parabola opens upward, and its vertex is the lowest point on the parabola, as shown in Figure 3.2. The y-coordinate of the vertex is the minimum value.

If $a < 0$, the parabola opens downward, and its vertex is the maximum point on the parabola, as shown in Figure 3.3. The y-coordinate of the vertex is the maximum value.

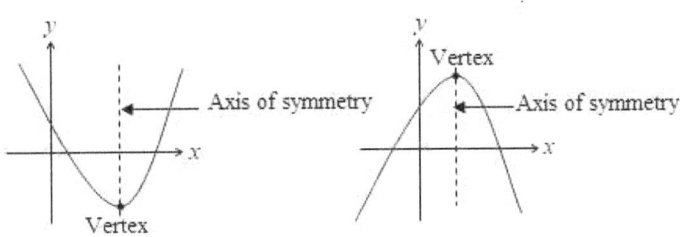

Figure 3.2   Figure 3.3

The graph of the quadratic function $y = ax^2 + bx + c$ intersects the y-axis at $(0, c)$. Substituting c for y, we have

$$c = ax^2 + bx + c$$

$$0 = ax^2 + bx$$

48 CHAPTER 3   Quadratic Functions

$$0 = x(ax + b)$$

Setting each factor to 0 and solving the linear equation for x, we get $x = 0$ or $x = -b/a$.

The x-coordinate of the midpoint of the line joining points $(0, 0)$ and $(-b/a, 0)$ is

$$\frac{0 + (-b/a)}{2} = -\frac{b}{2a}$$

Since the axis of symmetry passes through this point its equation is $x = -b/2a$. The y-coordinate of the vertex is obtained by substituting $-b/2a$ for x in the original equation.

Examples

1. Find the minimum value of the function $y = x^2 + 6x + 14$. The x-coordinate of the vertex is given by

$$x = -\frac{b}{2a}$$

Here $a = 1$ and $b = 6$.

$$x = -\frac{6}{2(1)} = -3$$

Substitute $x = -3$ into $y = x^2 + 6x + 14$

$$y = (-3)^2 + 6(-3) + 14$$

$$= 5$$

Because a is positive the function has a minimum value of 5, and this occur when $x = -3$.

2. Find the maximum value of the function $y = 3 - 10x - 2x^2$.

The x-coordinate of the vertex is

$$x = -\frac{(-10)}{2(-2)} = -\frac{5}{2}.$$

The y-coordinate of the vertex is

Sketching Graphs of Quadratic Functions

$$y = 3 - 10\left(-\tfrac{5}{2}\right) - 2\left(-\tfrac{5}{2}\right)^2$$

$$= 3 + 25 - \tfrac{25}{2}$$

$$= \tfrac{31}{2}$$

Because a is negative the function has a maximum value of 31/2, and this occur when $x = -5/2$.

3. Sketch the graph of the following functions.

(a) $y = x^2 - 4x - 12$

First, find the x- and y- intersects of the graph.

To find the y-intersect we substitute 0 for x giving $y = -12$

The y-intersect is $(0, -12)$.

To find the x-intersect(s) substitute 0 for y.

$$x^2 - 4x - 12 = 0$$

$$(x - 6)(x + 2) = 0$$

$$x = 6 \quad \text{or} \quad x = -2$$

The x-intersects are $(6, 0)$ and $(-2, 0)$.

Next, find the vertex of the curve.

The x-coordinate of the vertex is

$$x = -\tfrac{(-4)}{2(1)} = 2.$$

The y-coordinate of the vertex is $y = 2^2 - 4(2) - 12 = -16$.

Since a is positive, the graph opens upward. So the graph has a minimum point at $(2, -16)$.

Finally, plot these points and draw a smooth curve through the points. A sketch of the graph is shown in Figure 3.4.

50 CHAPTER 3 Quadratic Functions

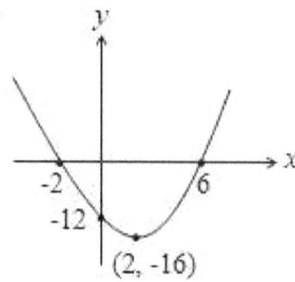

Figure 3.4

(b) $y = 15 - 2x - x^2$

Substituting $x = 0$ into $y = 15 - 2x - x^2$ gives $y = 15$.

Also, substituting 0 for y, we have

$$15 - 2x - x^2 = 0$$

$$(5 + x)(3 - x) = 0$$

$$x = -5 \text{ or } x = 3$$

The x-coordinate of the vertex is

$$x = -\frac{(-2)}{2(-1)} = -1.$$

The y-coordinate of the vertex is $y = 15 - 2(-1) - (-1)^2 = 16$.

Since a is negative, the graph opens downward and has a maximum point at $(-1, 16)$. A sketch of the graph is shown in Figure 3.5.

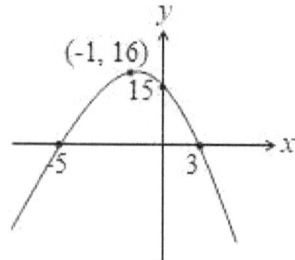

Figure 3.5

Sketching Graphs of Quadratic Functions

Finding the Range of a Quadratic Function

To find the range of a quadratic function, you need to find the lowest value or the highest value of the function.

Examples

Find the range of the following functions.

1. $f(x) = 2x^2 + 8x - 5$

    First, we find the vertex of the parabola.

    The x-coordinate of the vertex is

    $$x = -\frac{b}{2a} = -\frac{8}{2(2)} = -2$$

    and the y-coordinate is

    $$y = 2(-2)^2 + 8(-2) - 5 = -13.$$

    Because the coefficient of x^2 is positive, the parabola opens upward. So $(-2, -13)$ is the lowest point. Since the parabola extend upward, the range is all real numbers greater than or equal to -13, i.e. $y \geq -13$.

2. $f(x) = -3x^2 + 12x + 7$

    The x-coordinate of the vertex is

    $$x = -\frac{12}{2(-3)} = 2.$$

    and the y-coordinate is

    $$y = -3(2^2) + 12(2) + 7 = 19.$$

    Because the coefficient of x^2 is negative, the parabola opens downward. So $(2, 19)$ is the highest point. Since the parabola extend downward, the range is all real numbers less than or equal to 19, i.e. $y \leq 19$.

Exercise 3.4

1. Sketch the graph of each of the following quadratic functions.

    (a) $y = x^2 + 6x + 5$ (b) $y = 4 - 3x - x^2$

(c) $y = x^2 + 3x - 10$ (d) $y = x^2 - 2x - 15$

(e) $y = -x^2 + 6x - 8$ (f) $y = x^2 + 6x + 8$

(g) $y = 12 + x - x^2$ (h) $y = 6 + x - 2x^2$

(i) $y = 3x^2 + 10x + 3$ (j) $y = -2x^2 - 3x - 1$

(k) $y = 2x^2 + x - 1$ (l) $y = 2 - 5x - 3x^2$

2. Sketch the graph of each function, and use it to find the range of the function.

(a) $y = x^2 - 2x - 3$ (b) $y = 4 - 6x - x^2$

(c) $y = 3x^2 + 6x + 1$ (d) $y = -2x^2 + 12x + 3$

(e) $y = x^2 + x - 6$ (f) $y = 2x^2 - x - 6$

(g) $y = 4 - 3x - x^2$ (h) $y = 3 - 5x - 2x^2$

3.5 Solving Quadratic Inequalities

An expression such as $ax^2 + bx + c < 0$, where a, b and c are real numbers and $a \neq 0$ is called a quadratic inequality. The inequality symbol $<$ can be replaced by the symbol $>$, $\leq$ or $\geq$. Quadratic equations can be solved by the following method.

Algebraic Solution of Quadratic Inequalities

Examples

Solve the following inequalities.

1. $x^2 - 4x - 32 < 0$

First, find the solution of the corresponding quadratic equation.

$$x^2 - 4x - 32 = 0$$

$$(x - 8)(x + 4) = 0$$

$$x = 8 \quad \text{or} \quad x = -4$$

Solving Quadratic Inequalities

The zeros divide the number line into three intervals:

$$x < -4 \qquad -4 < x < 8 \qquad \text{and} \qquad x > 8$$

Next, we pick a number in each interval to determine whether the value of the function is negative or positive.

Intervals	Number chosen	$x^2 - 4x - 32$
$x < -4$	$x = -5$	$25 + 20 - 32 = 13$
$-4 < x < 8$	$x = 1$	$1 - 4 - 32 = -35$
$x > 8$	$x = 9$	$81 - 36 - 32 = 13$

If you get a negative value, all the values in the interval will be negative. Similarly, if you get a positive value, all values in the interval will be positive. You can see that the function is less than zero between −4 and 8. Hence, the solution is $-4 < x < 8$.

2. $(x - 4)(x + 8) \geq 13$

First, rewrite the inequality in the general form.

$$x^2 + 4x - 32 \geq 13$$

$$x^2 + 4x - 45 \geq 0$$

Next solve the corresponding quadratic equation.

$$x^2 + 4x - 45 = 0$$

$$(x - 5)(x + 9) = 0$$

$$x = 5 \quad \text{or} \quad x = -9$$

The zeros divide the number line into the following three intervals:

$$x < -9 \qquad -9 < x < 5 \qquad \text{and} \qquad x > 5$$

We pick a number in each interval to determine the value of the function.

Intervals	Number chosen	$x^2 + 4x - 45$
$x < -9$	$x = -10$	$100 - 40 - 45 = 15$
$-9 < x < 5$	$x = 1$	$1 + 4 - 45 = -40$
$x > 5$	$x = 6$	$36 + 24 - 45 = 15$

The function is more than 0 in the interval $x < -9$ and $x > 5$. The value of the function is 0 when $x = -9$ and $x = 5$. Hence, the solution is $x \leq -9$ or $x \geq 5$.

Exercise 3.5(a)

Solve the following inequalities.

1. $x^2 - 3x - 4 > 0$
2. $x^2 - 2x - 8 < 0$
3. $x^2 - x - 12 \leq 0$
4. $x^2 + 7x + 12 \geq 0$
5. $x^2 - 3x - 10 < 0$
6. $x^2 + 6x - 27 > 0$
7. $(x + 3)(x - 4) \leq 0$
8. $x^2 - 3x > 18$
9. $2x^2 + 7x + 3 < 0$
10. $3x^2 + 8x + 5 \geq 0$
11. $1 - x - 2x^2 \leq 0$
12. $6 + 7x \geq 3x^2$

Graphical Solution of Quadratic Inequalities

To solve a quadratic inequality such as $3x^2 + 2x + 5 < 0$, we sketch the graph of $y = 3x^2 + 2x + 5$ and then determine the interval where the graph is below the axis.

Examples

Solve the following inequalities.

1. $x^2 + 2x - 15 < 0$

   First, sketch a graph of $y = x^2 + 2x - 15$. A graph of the function is shown in Figure 3.6.

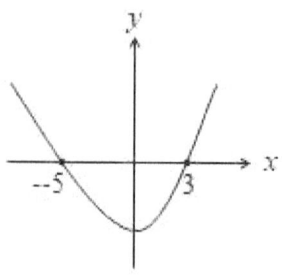

Figure 3.6

Next, we need to find where the graph is below the x-axis. You may observe that the graph is below the x-axis between -5 and 3. Hence, the solution is $-5 < x < 3$.

2. $x^2 - 3x - 20 \geq 8$

First, rewrite the inequality in the general form:

$$x^2 - 3x - 28 \geq 0$$

Next, sketch the graph of $y = x^2 - 3x - 28$ and determine where the graph is above the x-axis. A graph of the function is shown in Figure 3.7.

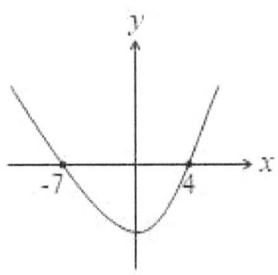

Figure 3.7

The graph is above the x-axis, when x is less than -7 or when x is greater than 4. Hence, the solution is $x \leq -7$ or $x \geq 4$.

Exercise 3.5(b)

Solve the following inequalities.

1. $x^2 + x - 12 < 0$
2. $x^2 + 3x - 10 > 0$
3. $x^2 + 4x - 12 > 0$
4. $x^2 + 2x - 15 < 0$
5. $x^2 - 5x + 6 \geq 0$
6. $x^2 + 7x + 10 \leq 0$
7. $2x^2 + x - 6 \leq 0$
8. $3x^2 - 10x - 8 \leq 0$
9. $4x^2 + x - 3 < 0$
10. $x(x + 5) \geq 24$

56 CHAPTER 3   Quadratic Functions

Review Exercise 3

1. Solve the following quadratic equations.

   (a) $x^2 + 3x - 10 = 0$

   (b) $x^2 - 8x - 20 = 0$

   (c) $x^2 - 9x + 20 = 0$

   (d) $x^2 - 2x - 10 = 0$

   (e) $2 + 6x - 5x^2 = 0$

   (f) $x^2 = 10x - 21$

   (g) $x^2 - 1 = 12x + 12$

   (h) $3x^2 - 2x = 9$

   (i) $5x^2 + 7 = 15x$

   (j) $2x^2 - x - 3 = 0$

   (k) $(2x + 1)^2 = 49$

   (l) $4x^2 + 3x = 6$

2. Find the nature of the root of each of the following equations.

   (a) $x^2 - 2x + 1 = 0$

   (b) $2x^2 - 3x - 2 = 0$

   (c) $5x^2 + 7x + 3 = 0$

   (d) $3x^2 - 2x - 5 = 0$

3. Find the values of k for which the equation $x^2 - (k + 1)x + 2k - 1 = 0$ has equal roots.

4. Find the range of values of k for which the equation $kx^2 + 6x + 3 = 0$ has two distinct real roots.

5. Find the smallest possible integral value of k such that $3x^2 - 4x + k = 0$ will have imaginary roots.

6. Find the range of values of k for which the expression $3x^2 + 2x + k$ is always positive for all real values of x.

7. Find the minimum or the maximum value of the following expression.

   (a) $2x^2 + 6x + 2$

   (b) $-x^2 + 6x - 10$

   (c) $x^2 - 4x + 7$

   (d) $x^2 + 6x + 5$

   (e) $4 - 8x - x^2$

   (f) $3x^2 + 6x + 1$

   (g) $12 + 5x - 3x^2$

   (h) $2x^2 - 7x + 5$

   (i) $2 - 2x - 6x^2$

8. Sketch the graph of the following functions.

   (a) $y = x^2 - 4x - 5$

   (b) $y = 3 + 2x - x^2$

   (c) $y = 2x^2 - 5x - 3$

   (d) $y = 5 + x - 4x^2$

9. Sketch the graph of each function, and use it to find the range of the function.

   (a) $y = x^2 - 8x + 15$

   (b) $y = 5 - 6x - 2x^2$

   (c) $y = 3x^2 - 6x + 4$

10. Solve the following inequalities.

   (a) $x^2 + 4x - 5 \geq 0$

   (b) $x^2 - 6x - 7 < 0$

   (c) $x^2 + 3x - 10 \leq 0$

   (d) $8x - x^2 > 12$

   (e) $3x^2 + 2x - 8 \leq 0$

   (f) $2x^2 - 3x \geq 20$

4 Indices and Surds

4.1 Indices

If n is a positive integer and a a real number, then the product of n factors of a is written briefly as a^n, read " a raised to the power n."

For example, $2 \cdot 2 \cdot 2 \cdot 2 \cdot 2 \cdot 2 \cdot 2 = 2^7$

The expression a^n is written in index form. The number a is called the base and n is the index. The plural of index is indices.

Basic Laws of Indices

Consider the product $a^2 \times a^3$

The expression a^2 represents two factors of a and a^3 represents three factors of a. To find the product work as follows.

$$a^2 \times a^3 = (a \cdot a) \cdot (a \cdot a \cdot a)$$
$$= a \cdot a \cdot a \cdot a \cdot a$$
$$= a^5$$

You can see that, the product of two expressions in index form is obtained by adding all index. In, general, if a is non-zero real number, and m and n are positive integers then

$$a^m \times a^n = a^{m+n}$$

In a similar way, you can verify that:

$$a^m \div a^n = a^{m-n}$$
$$(a^m)^n = a^{m \cdot n}$$

Examples

Simplify the following.

1. $5^3 \times 5^4$

$$5^3 \times 5^4 = 5^{(3+4)}$$
$$= 5^7$$

60 CHAPTER 4 Indices and Surds

2. $3^{10} \div 3^6$

$$3^{10} \div 3^6 = 3^{(10-6)}$$
$$= 3^4$$

3. $(2^3)^4 = 2^{(3 \cdot 4)}$
$$= 2^{12}$$

Example 4.1(a)

1. Simplify the following. Leave the answer in index form.

   (a) $4^2 \times 4^3$ (b) $3^2 \times 3^4 \times 3^3$ (c) $2 \times 2^3 \times 2^4$

   (d) $5^6 \div 5^2$ (e) $7^8 \div 7^5$ (f) $4^3 \div 4$

   (g) $(5^3)^2$ (h) $(3^3)^5$ (i) $(7^2)^4$

2. Simplify the following. Leave the answer in index form.

   (a) $\dfrac{2^3 \times 2^5}{2^4}$ (b) $\dfrac{3^6 \times 3^7}{3^2 \times 3^8}$ (c) $\dfrac{(5^2)^6 \times 5^3}{5^8}$

3. Simplify the following.

   (a) $\dfrac{8a^6 b^4}{4a^2 b^3}$ (b) $\dfrac{12x^5 y^7}{4x^2 y^5}$ (c) $\dfrac{5x^6 y^7 z^4}{15x^4 y^3 z^2}$

4. Simplify the following.

   (a) $5a^2 \times 3a^3$ (b) $3a^2 b \times 2a^3 b^2$

   (c) $(3x^2 y^3)^2$ (d) $(2x^2)^3 \times (3x^3)^2$

   (e) $(3x^2 y)^2 \times (2xy^3)^3$ (f) $\dfrac{(4x^2 y^4)^2 \times (2xy^2)^3}{32 x^5 y^8}$

The definition of an index can be extended to include zero, negative and fractional index.

Zero Index

For any real number a, where $a \neq 0$.

$$a^0 \times a^n = a^n$$

$$a^0 = \frac{a^n}{a^n}$$

$$a^0 = 1$$

Any number raised to the power 0 is equal to 1.

Negative Index

For any real number a, where $a \neq 0$.

$$a^{-n} \times a^n = a^0$$

$$a^{-n} = \frac{a^0}{a^n}$$

$$a^{-n} = \frac{1}{a^n}$$

If n is an integer, then a^{-n} is defined as the reciprocal of a^n.

Fractional Index

For any nonzero real number a.

$$a^{1/3} \cdot a^{1/3} \cdot a^{1/3} = a^{(1+1+1)/3}$$

$$= a$$

Therefore,

$$a^{1/3} = \sqrt[3]{a}.$$

In general, if a is any real number and n is a positive integer then

$$a^{1/n} = \sqrt[n]{a}$$

Also

$$a^{2/3} \cdot a^{2/3} \cdot a^{2/3} = a^{(2+2+2)/3}$$

$$= a^2$$

Therefore,

$$a^{2/3} = \sqrt[3]{a^2}.$$

so, $a^{2/3} = (a^{1/3})^2 = (\sqrt[3]{a})^2$

In general, if a is any real number, and m and n are positive integers greater than 1, then

$$a^{m/n} = \sqrt[n]{a^m} = (\sqrt[n]{a})^m.$$

The laws of indices are summarized below for easy reference.

1. $a^m \times a^n = a^{m+n}$

2. $a^m \div a^n = a^{m-n}$

3. $(a^m)^n = a^{mn}$

4. $a^{1/n} = \sqrt[n]{a}$

5. $a^{m/n} = \sqrt[n]{a^m} = (\sqrt[n]{a})^m$

6. $a^{-n} = \dfrac{1}{a^n}$

7. $a^0 = 1$

Examples

1. Find the values of the following expressions.

(a) $\dfrac{(2^3)^4 \times (2^{-2})^5}{(3^2)^4 \div (3^5)^2}$

$$\dfrac{(2^3)^4 \times (2^{-2})^5}{(3^2)^4 \div (3^5)^2} = \dfrac{2^{12} \times 2^{-10}}{3^8 \div 3^{10}}$$

$$= \dfrac{2^2}{3^{-2}}$$

$$= 2^2 \times 3^2$$

$$= 36$$

(b) $\dfrac{\sqrt[3]{5^4}\times(5^{-4})^{1/3}}{(2^5)^{-3}\times(2^4)^3}$

$$\dfrac{\sqrt[3]{5^4}\times(5^{-4})^{1/3}}{(2^5)^{-3}\times(2^4)^3} = \dfrac{5^{4/3}\times 5^{-4/3}}{2^{-15}\times 2^{12}}$$

$$= \dfrac{5^0}{2^{-3}}$$

$$= 2^3$$

$$= 8$$

2. Simplify $\dfrac{\left(x^{-1/4}y^{-1/3}\right)^6}{x^{1/2}y^{-8}}$

$$\dfrac{\left(x^{-1/4}y^{-1/3}\right)^6}{x^{1/2}y^{-8}} = \dfrac{x^{-3/2}y^{-2}}{x^{1/2}y^{-8}}$$

$$= x^{-2}y^6$$

$$= \dfrac{y^6}{x^2}$$

Exercise 4.1(b)

1. Find the value of each of the following expressions.

(a) $36^{\frac{1}{2}}$ (b) $125^{\frac{1}{3}}$ (c) $81^{\frac{1}{4}}$

(d) $32^{\frac{3}{5}}$ (e) $64^{\frac{2}{3}}$ (f) $8^{\frac{4}{3}}$

(g) $(-7)^0$ (h) $25^{-\frac{1}{2}}$ (i) $625^{-\frac{1}{4}}$

(j) $128^{-\frac{5}{7}}$ (k) $16^{-\frac{3}{4}}$ (l) $81^{\frac{3}{4}}$

2. Simplify the following.

(a) $\left(\dfrac{1}{32}\right)^{-3/5}$ (b) $\left(\dfrac{1}{27}\right)^{-2/3}$ (c) $\left(\dfrac{1}{16}\right)^{-3/4}$

(d) $\left(\dfrac{64}{27}\right)^{-2/3}$ (e) $\left(\dfrac{16}{81}\right)^{-3/4}$ (f) $\left(\dfrac{125}{216}\right)^{-2/3}$

(g) $\left(1\frac{7}{9}\right)^{-3/2}$ (h) $\left(3\frac{3}{8}\right)^{2/3}$ (i) $(0.0081)^{-1/4}$

3. Evaluate:

(a) $2^{-1} \cdot 3^2 \cdot 9^0$ (b) $20^{1/2} \cdot 5^{1/2}$ (c) $27^{1/4} \cdot 3^{1/4}$

(d) $243^{1/2} \cdot 3^{-1/2}$ (e) $\dfrac{27^{1/3} \cdot 125^{1/2}}{5^{1/2}}$ (f) $\dfrac{3^{1/3} \cdot 3^0 \cdot 9^{1/3}}{27^{1/3}}$

(g) $\dfrac{8^{1/3} \cdot 16^{1/3}}{32^{-1/3}}$ (h) $\dfrac{4^{1/3} \cdot 8^{-1/2}}{2^{-1/2} \cdot 4^{-2/3}}$ (i) $\dfrac{75^{1/3} \cdot 15^{1/3}}{9^{1/3}}$

4. Simplify:

(a) $\dfrac{(x^{1/2}y^{-1})^3}{x^{-1/2}y}$ (b) $\dfrac{(x^{-1}y^3)^4}{(x^2y^{-6})^{-2}}$ (c) $\left(\dfrac{8x^{-6}y^3}{z^{-9}}\right)^{1/3}$

(d) $\left(\dfrac{x^{3/4}y^{-1/2}}{z^{1/4}}\right)^4$ (e) $\dfrac{(x^2y^6)^{1/5}}{\sqrt[5]{x^7y}}$ (f) $\dfrac{\sqrt{x^3} \cdot x^5}{\sqrt{x^5}}$

Exponential Equation

An equation that contains an expression in index form is called an exponential equation. Examples include $2^x = 8$ and $125^{2x-3} = 5$.

Solving Exponential Equations

Examples

Solve the following equations.

1. $3^{2x+1} = 243$

$\qquad 3^{2x+1} = 243 \qquad$ Original equation

$\qquad 3^{2x+1} = 3^5 \qquad$ Rewrite 243 as 3^5

$\qquad 2x + 1 = 5 \qquad$ Equate the index

$\qquad 2x = 4 \qquad$ Subtract 1 from each side

$\qquad x = 2 \qquad$ Divide each side by 2

2. $\dfrac{8^{3x-2}}{16^{2x}} = \dfrac{32^{x-3}}{4^{x-2}}$

$$\dfrac{8^{3x-2}}{16^{2x}} = \dfrac{32^{x-3}}{4^{x-2}} \qquad \text{Original equation}$$

$$\dfrac{2^{9x-6}}{2^{8x}} = \dfrac{2^{5x-15}}{2^{2x-4}} \qquad \text{Rewrite the expression.}$$

$$2^{x-6} = 2^{3x-11} \qquad \text{Simplify}$$

$$x - 6 = 3x - 11 \qquad \text{Equate the index}$$

$$-2x = -5 \qquad \text{Simplify}$$

$$x = \dfrac{5}{2} \qquad \text{Divide each side by } -2$$

Exercise 4.1(c)

1. Solve each of the following equation.

    (a) $8^x = 64$

    (b) $9^x = 27$

    (c) $128^x = 32$

    (d) $3^{2x-1} = 27$

    (e) $2^{x-1} = \dfrac{1}{8}$

    (f) $8^{x-1} = 16$

    (g) $(x-1)^3 = 64$

    (h) $x^{\frac{3}{4}} = 27$

    (i) $\left(\dfrac{9}{25}\right)^{-x} = \left(\dfrac{5}{3}\right)^{3-x}$

    (j) $9^{x-2} = 3^{3x-1}$

    (k) $4^{2x} = \dfrac{1}{2}(8)^x$

    (l) $3^{\sqrt{x}} = 27^x$

2. Solve the following equations.

    (a) $\dfrac{25^x}{5^{5-x}} = \dfrac{5^{4x}}{125^{x-3}}$

    (b) $\dfrac{3^{5x-2}}{9^x} = \dfrac{27^{x-1}}{9^{3-x}}$

    (c) $\dfrac{36^{-2x}}{6^{3-5x}} = \dfrac{36^{3x-2}}{216^{2-x}}$

    (d) $\dfrac{8^{3-x}}{4^{2x}} = \dfrac{2^{3x-5}}{4^{3x}}$

4.2 Surds

The positive square root of a number a, denoted by $\sqrt{a}$ is a positive number whose square is a. For example, because $3^2 = 9$, $\sqrt{9} = 3$.

Square roots of perfect squares, such as 4, 25 and 144 are rational numbers. Square roots of numbers that are not perfect squares are irrational numbers. Examples include $\sqrt{2}$, $\sqrt{3}$ and $\sqrt{25}$. These numbers are generally called surds.

Simplifying Surds

A surd is in simplified form when the radicand has no perfect square factors other than 1. The following properties help us simplify surds.

Let a be a non-negative integer, then

1. $\sqrt{a^2} = \left(\sqrt{a}\right)^2 = a$

2. $\sqrt{ab} = \sqrt{a} \times \sqrt{b}$

3. $\sqrt{\dfrac{a}{b}} = \dfrac{\sqrt{a}}{\sqrt{b}}$

To simplify a surd, find the largest perfect square factor and then use the fact that $\sqrt{a^2 b} = \sqrt{a^2} \times \sqrt{b} = a\sqrt{b}$.

Examples

Simplify the following surds.

1. $\sqrt{18}$

$$\sqrt{18} = \sqrt{9 \cdot 2}$$
$$= \sqrt{9} \cdot \sqrt{2}$$
$$= 3\sqrt{2}$$

2. $\sqrt{108}$

$$\sqrt{108} = \sqrt{36 \cdot 2}$$
$$= \sqrt{36} \cdot \sqrt{2}$$
$$= 6\sqrt{2}$$

Exercise 4.2(a)

Simplify the following surds.

1. $\sqrt{20}$ 2. $\sqrt{28}$ 3. $\sqrt{27}$ 4. $\sqrt{40}$

5. $\sqrt{45}$ 6. $\sqrt{48}$ 7. $\sqrt{50}$ 8. $\sqrt{54}$

9. $\sqrt{60}$ 10. $\sqrt{72}$ 11. $\sqrt{75}$ 12. $\sqrt{80}$

13. $\sqrt{96}$ 14. $\sqrt{98}$ 15. $\sqrt{147}$ 16. $\sqrt{150}$

17. $\sqrt{162}$ 18. $\sqrt{180}$ 19. $\sqrt{192}$ 20. $\sqrt{245}$

Adding and Subtracting Surds

To add (or subtract) surds, we apply the distributive property and then combine the coefficients as the following examples illustrate.

Examples

Add:

1. $2\sqrt{3} + 4\sqrt{3}$

$$2\sqrt{3} + 4\sqrt{3} = (2 + 4)\sqrt{3}$$
$$= 6\sqrt{3}$$

In practice, it is not necessary to show the intermediate step.

2. $5\sqrt{2} - 4\sqrt{2}$

$$5\sqrt{2} - 4\sqrt{2} = \sqrt{2}$$

CHAPTER 4 Indices and Surds

Often you need to rewrite the surds in simplest form before you add or subtract.

Examples

1. $\sqrt{24} + \sqrt{96}$

$$\sqrt{24} + \sqrt{96} = 2\sqrt{6} + 4\sqrt{6}$$
$$= 6\sqrt{6}$$

2. $2\sqrt{45} - 3\sqrt{20}$

$$2\sqrt{45} - 3\sqrt{20} = 6\sqrt{5} - 6\sqrt{5} = 0$$

Exercise 4.2(b)

Workout:

1. $5\sqrt{3} + 2\sqrt{3}$
2. $3\sqrt{2} + 5\sqrt{2}$
3. $10\sqrt{5} - 3\sqrt{5}$
4. $6\sqrt{7} - 4\sqrt{7}$
5. $2\sqrt{3} - 5\sqrt{3}$
6. $2\sqrt{5} - 4\sqrt{5}$
7. $2\sqrt{3} + \sqrt{12}$
8. $5\sqrt{2} + \sqrt{18}$
9. $3\sqrt{12} - \sqrt{48}$
10. $2\sqrt{45} - 2\sqrt{20}$
11. $2\sqrt{2} + \sqrt{2} + 3\sqrt{2}$
12. $3\sqrt{3} + 2\sqrt{3} + \sqrt{3}$
13. $5\sqrt{5} - 2\sqrt{5} + \sqrt{5}$
14. $2\sqrt{7} + 3\sqrt{7} - 8\sqrt{7}$
15. $\sqrt{50} + \sqrt{32} - \sqrt{8}$
16. $5\sqrt{8} + 3\sqrt{18} - 4\sqrt{32}$

Multiplying Surds

Surds are multiplied in much the same way that algebraic expressions are multiplied.

Examples

Multiple:

1. $\sqrt{2} \times \sqrt{10}$

$$\sqrt{2} \times \sqrt{10} = \sqrt{2 \cdot 10}$$
$$= \sqrt{20}$$
$$= 2\sqrt{5}$$

2. $2\sqrt{3} \times 3\sqrt{6}$

$$2\sqrt{3} \times 3\sqrt{6} = 2 \cdot 3 \cdot \sqrt{3} \cdot \sqrt{6}$$
$$= 6\sqrt{18}$$
$$= 6 \cdot 3\sqrt{2}$$
$$= 18\sqrt{2}$$

3. $\sqrt{32} \times \sqrt{45}$

$$\sqrt{32} \times \sqrt{45} = 4\sqrt{2} \times 3\sqrt{5}$$
$$= 12\sqrt{10}$$

Exercise 4.2(c)

Workout:

1. $\sqrt{2} \times \sqrt{6}$
2. $\sqrt{5} \times \sqrt{15}$
3. $5\sqrt{3} \times 2\sqrt{5}$
4. $3\sqrt{2} \times 2\sqrt{5}$
5. $\sqrt{8} \times \sqrt{10}$
6. $\sqrt{15} \times \sqrt{27}$
7. $\sqrt{50} \times \sqrt{30}$
8. $\sqrt{98} \times \sqrt{75}$
9. $\sqrt{3} \times \sqrt{2} \times \sqrt{10}$
10. $\sqrt{6} \times 2\sqrt{2} \times \sqrt{15}$
11. $\sqrt{30} \times \sqrt{2} \times \sqrt{27}$
12. $\sqrt{72} \times \sqrt{75} \times \sqrt{18}$

Rationalizing the Denominator

A fraction whose denominator is a surd can be rewritten in an equivalent form without a surd in the denominator. The procedure for removing the surds from the denominator is called rationalizing the denominator.

Examples

Rationalize the denominator of the following expressions.

1. $\dfrac{1}{\sqrt{3}}$

   Multiply both the numerator and denominator by $\sqrt{3}$.

   $$\dfrac{1}{\sqrt{3}} = \dfrac{1}{\sqrt{3}} \times \dfrac{\sqrt{3}}{\sqrt{3}}$$

   $$= \dfrac{\sqrt{3}}{3}$$

2. $\dfrac{8}{\sqrt{2}}$

   $$\dfrac{8}{\sqrt{2}} = \dfrac{8}{\sqrt{2}} \times \dfrac{\sqrt{2}}{\sqrt{2}}$$

   $$= \dfrac{8\sqrt{2}}{2}$$

   $$= 4\sqrt{2}$$

3. $\dfrac{9}{2\sqrt{3}}$

   $$\dfrac{9}{2\sqrt{3}} = \dfrac{9}{2\sqrt{3}} \times \dfrac{2\sqrt{3}}{2\sqrt{3}}$$

   $$= \dfrac{18\sqrt{3}}{12}$$

   $$= \dfrac{3\sqrt{3}}{2}$$

   You can obtain the same result as follows.

$$\frac{9}{2\sqrt{3}} = \frac{9}{2\sqrt{3}} \times \frac{\sqrt{3}}{\sqrt{3}}$$

$$= \frac{9\sqrt{3}}{6}$$

$$= \frac{3\sqrt{3}}{2}$$

Exercise 4.2(d)

Rationalize the denominator:

1. $\frac{2}{\sqrt{3}}$
2. $\frac{10}{\sqrt{5}}$
3. $\frac{9}{\sqrt{3}}$
4. $\frac{15}{\sqrt{10}}$

5. $\frac{18}{\sqrt{12}}$
6. $\frac{4\sqrt{3}}{\sqrt{6}}$
7. $\frac{3\sqrt{5}}{\sqrt{6}}$
8. $\frac{12\sqrt{2}}{\sqrt{6}}$

9. $\frac{8}{\sqrt{32}}$
10. $\frac{3}{4\sqrt{2}}$
11. $\frac{5}{3\sqrt{10}}$
12. $\frac{14}{5\sqrt{7}}$

Expanding Expressions containing Surds

The distributive property can be applied in multiplying expressions with surds, as illustrated in the following examples.

Examples

Expand the following expressions.

1. $3\sqrt{2}(\sqrt{3} + 2\sqrt{2})$

$$3\sqrt{2}(\sqrt{3} + 2\sqrt{2}) = 3\sqrt{2} \cdot \sqrt{3} + 3\sqrt{2} \cdot 2\sqrt{2}$$

$$= 3\sqrt{6} + 12$$

2. $(2 + \sqrt{3})(4 - \sqrt{3})$

$$(2 + \sqrt{3})(4 - \sqrt{3}) = 2 \cdot 4 - 2 \cdot \sqrt{3} + \sqrt{3} \cdot 4 - \sqrt{3} \cdot \sqrt{3}$$

$$= 8 - 2\sqrt{3} + 4\sqrt{3} - 3$$

$$= 5 + 2\sqrt{3}$$

3. $(3\sqrt{3} - 5)(3\sqrt{3} + 5)$

$$(3\sqrt{3} - 5)(3\sqrt{3} + 5) = 3\sqrt{3} \cdot 3\sqrt{3} + 3\sqrt{3} \cdot 5 - 5 \cdot 3\sqrt{3} - 5 \cdot 5$$
$$= 9 \cdot 3 + 15\sqrt{3} - 15\sqrt{3} - 25$$
$$= 27 - 25$$
$$= 2$$

4. $(2\sqrt{5} - 3)^2$

$$(2\sqrt{5} - 3)^2 = (2\sqrt{5} - 3)(2\sqrt{5} - 3)$$
$$= 2\sqrt{5} \cdot 2\sqrt{5} - 2\sqrt{5} \cdot 3 - 3 \cdot 2\sqrt{5} + 3 \cdot 3$$
$$= 20 - 6\sqrt{5} - 6\sqrt{5} + 9$$
$$= 29 - 12\sqrt{5}$$

Exercise 4.2(e)

Simplify each expression.

1. $\sqrt{3}(\sqrt{5} - \sqrt{3})$
2. $\sqrt{2}(3\sqrt{15} - 4\sqrt{3})$
3. $\sqrt{6}(\sqrt{6} + \sqrt{5})$
4. $\sqrt{7}(2\sqrt{3} + 3\sqrt{7})$
5. $(\sqrt{3} + 5)(\sqrt{3} + 2)$
6. $(\sqrt{5} - 2)(\sqrt{5} - 1)$
7. $(3\sqrt{2} - \sqrt{3})(2\sqrt{3} + \sqrt{2})$
8. $(2\sqrt{2} + \sqrt{3})(\sqrt{12} - \sqrt{8})$
9. $(\sqrt{8} - 2)(\sqrt{8} + 2)$
10. $(2\sqrt{3} + 4\sqrt{2})(2\sqrt{3} - 4\sqrt{2})$
11. $(2 + 3\sqrt{2})^2$
12. $(3\sqrt{2} - 2\sqrt{3})^2$

Fractions whose denominators are sum or difference of surds

Whenever expressions of the form $\sqrt{a} + \sqrt{b}$ and $\sqrt{a} - \sqrt{b}$ and $a - \sqrt{b}$ and $a + \sqrt{b}$ are multiplied the product is always a rational number. For example,

$$(\sqrt{7} + \sqrt{3})(\sqrt{7} - \sqrt{3}) = 7 - 3 = 4.$$

The expressions

$$\sqrt{7} + \sqrt{3} \text{ and } \sqrt{7} - \sqrt{3}$$

are conjugates.

The method used in rationalizing fractions whose denominators are sum or difference of surds is illustrated in the following example.

Examples

Rationalize the denominator:

1. $\dfrac{2\sqrt{3}}{2 + \sqrt{3}}$

   We multiply both the numerator and denominator by the conjugate of $2 + \sqrt{3}$.

   $$\frac{2\sqrt{3}}{2+\sqrt{3}} = \frac{2\sqrt{3}}{2+\sqrt{3}} \times \frac{2-\sqrt{3}}{2-\sqrt{3}}$$

   $$= \frac{2\sqrt{3}(2-\sqrt{3})}{(2+\sqrt{3})(2-\sqrt{3})}$$

   $$= 4\sqrt{3} - 6$$

2. $\dfrac{4\sqrt{5} + 2\sqrt{3}}{\sqrt{5} - \sqrt{3}}$

   $$\frac{4\sqrt{5}+2\sqrt{3}}{\sqrt{5}-\sqrt{3}} = \frac{4\sqrt{5}+2\sqrt{3}}{\sqrt{5}-\sqrt{3}} \times \frac{\sqrt{5}+\sqrt{3}}{\sqrt{5}+\sqrt{3}}$$

   $$= \frac{(4\sqrt{5}+2\sqrt{3})(\sqrt{5}+\sqrt{3})}{(\sqrt{5})^2 - (\sqrt{3})^2}$$

   $$= \frac{26 + 6\sqrt{15}}{2}$$

   $$= 13 + 3\sqrt{15}$$

CHAPTER 4 Indices and Surds

Exercise 4.2(f)

Rationalize the denominator:

1. $\dfrac{1}{\sqrt{2}+1}$

2. $\dfrac{1}{2-\sqrt{3}}$

3. $\dfrac{5}{4+\sqrt{6}}$

4. $\dfrac{3\sqrt{7}}{5-\sqrt{7}}$

5, $\dfrac{2\sqrt{6}}{2\sqrt{3}+1}$

6. $\dfrac{2\sqrt{3}-\sqrt{5}}{2\sqrt{3}+\sqrt{5}}$

7. $\dfrac{3-2\sqrt{2}}{3+2\sqrt{2}}$

8. $\dfrac{\sqrt{3}+\sqrt{5}}{\sqrt{3}-\sqrt{5}}$

9. $\dfrac{\sqrt{7}-3\sqrt{2}}{\sqrt{7}-\sqrt{2}}$

10. $\dfrac{\sqrt{6}+2\sqrt{2}}{\sqrt{6}-2\sqrt{2}}$

11. $\dfrac{\sqrt{5}+3}{4-\sqrt{20}}$

12. $\dfrac{\sqrt{3}+\sqrt{6}}{3\sqrt{2}-3}$

Review Exercise 4

1. Simplify:

   (a) $5^2 \times 5^4$

   (b) $3^8 \times 3^{-5}$

   (c) $2^3 \times 2^5$

   (d) $7^6 \div 7^3$

   (e) $5^2 \div 5^{-3}$

   (f) $3^{-7} \div 3^{-9}$

   (g) $(2^4)^2$

   (h) $(3^2)^3$

   (i) $(5^{-3})^{-2}$

2. Simplify:

   (a) $\dfrac{3^4 \times 3^6}{3^7}$

   (b) $\dfrac{7^{-3} \times 7^8}{7^5 \div 7^3}$

   (c) $\dfrac{5^9 \div 5^7}{5^3 \times 5^{-2}}$

3. Simplify:

   (a) $\dfrac{9a^5 b^7}{6a^3 b^5}$

   (b) $\dfrac{15x^{-3} y^8}{5x^{-7} y^6}$

   (c) $\dfrac{27z^3 y^{-6}}{18z^{-5} y^{-8}}$

4. Simplify:

   (a) $8x^3 \times 2x^4$

   (b) $2xy^3 \times 3x^2y^2$

   (c) $5x^2y^{-3} \times 3x^{-1}y^5$

5. Simplify:

   (a) $\left(\dfrac{x^3 y^{-6}}{z^{12}}\right)^{-2/3}$

   (b) $\left(\dfrac{x^{1/4} y^{-1/2}}{x^{1/2} y^{3/4}}\right)^4$

   (c) $\left(\dfrac{y^{-1}\sqrt[3]{x}}{\sqrt[3]{z^4}}\right)^3$

6. Simplify:

   (a) $\dfrac{(3^4)^5 \times (3^{-3})^7}{(2^3)^2 \div (2^4)^2}$

   (b) $\dfrac{(7^3)^2 \times (7^{-2})^3}{(5^4)^3 \div (5^3)^5}$

7. Solve:

   (a) $2^{x-4} = 8^2$

   (b) $3^{3x} = 9$

   (c) $4^{3x-1} = 64$

   (d) $5^{2-x} = 25$

   (e) $4^{x+3} = 32^x$

   (f) $9^{x-2} = 234^{x+1}$

8. Solve:

   (a) $\dfrac{81^{x+2}}{27^{x-3}} = \dfrac{3^{2-x}}{9^{2x}}$

   (b) $\dfrac{8^{2-3x}}{16^{3x}} = \dfrac{32^{3-x}}{4^{x+2}}$

9. Simplify:

   (a) $3\sqrt{6} - 8\sqrt{6}$

   (b) $2\sqrt{2} + 3\sqrt{2} - \sqrt{2}$

   (c) $\sqrt{5} - 3\sqrt{5} - 4\sqrt{5}$

   (d) $(\sqrt{3} + 2\sqrt{2})(\sqrt{3} + 5\sqrt{2})$

   (e) $(5\sqrt{3} - 2)^2$

   (f) $(2\sqrt{3} + \sqrt{2})(2\sqrt{3} - \sqrt{2})$

10. Rationalize the denominator:

(a) $\dfrac{1}{5\sqrt{2}}$ (b) $\dfrac{8}{3\sqrt{2}}$

(c) $\dfrac{8}{3\sqrt{18}}$ (d) $\dfrac{9}{4\sqrt{6}}$

(e) $\dfrac{3\sqrt{2}}{2\sqrt{2}-3}$ (f) $\dfrac{\sqrt{3}+2}{2\sqrt{3}-1}$

(g) $\dfrac{\sqrt{3}+\sqrt{2}}{\sqrt{3}-\sqrt{2}}$ (h) $\dfrac{2-\sqrt{3}}{\sqrt{6}+\sqrt{2}}$

5 Polynomials

A polynomial function of x is a term or a sum of terms, where each term is a whole number power of x multiplied by a constant. Examples include

$$5x^4 + 3x^2 + 7x,\ 4x^3 - 3x^2 + 2x - 5,\ 3x + 2,\ 6x \text{ and } 5$$

5.1 Dividing Polynomials

A polynomial can be divided by another polynomial by long division, as illustrated in the following example.

Examples

1. Divide $2x^3 + 5x^2 - 6x + 3$ by $x + 3$.

   Divide the first term, $2x^3$, by the first term in the divisor: $2x^3/x = 2x^2$. Then multiply $(x + 3)$ by $2x^2$. Subtract the answer from $(2x^3 + 5x^2)$ and bring down the next term. We repeat the process until we get a polynomial of degree less than that of the divisor.

$$
\begin{array}{r}
2x^2 - x - 3 \\
x+3\overline{)\,2x^3 + 5x^2 - 6x + 3} \\
\end{array}
$$

$2x^3 + 6x^2$	Multiply $x + 3$ by $2x^2$
$-x^2 - 6x$	Subtract and bring down $-6x$
$-x^2 - 3x$	Multiply $x + 3$ by $-x$
$-3x + 3$	Subtract and bring down 3
$-3x - 9$	Multiply $x + 3$ by -2
12	Subtract

   So the quotient when $2x^3 + 5x^2 - 6x + 3$ is divided by $(x + 3)$ is $2x^2 - x - 3$, and the remainder is 12.

2. Divide $x^3 - 8$ by $x + 2$.

   Any missing term of the dividend is written in using 0 for the coefficients.

CHAPTER 5 Polynomials

$$\begin{array}{r} x^2+2x+4 \\ x-2{\overline{\smash{\big)}\,x^3+0x^2+0x-8}} \end{array}$$

$\underline{x^3-2x^2}$ \qquad Multiply $x-2$ by x^2

$2x^2+0x$ \qquad Subtract and bring down $0x$

$\underline{2x^2-4x}$ \qquad Multiply $x-2$ by $2x$

$4x-8$ \qquad Subtract and bring down -8

$\underline{4x-8}$ \qquad Multiply $x-2$ by 4

0 \qquad Subtract

So the quotient when x^3-8 is divided by $(x-2)$ is x^2+2x+4, and the remainder is 0. This means that $(x-2)$ is a factor of x^3-8.

Exercise 5.1

1. Find the remainder in each of the following divisions.

   (a) $(x^3 - 3x^2 - 5x + 2) \div (x - 3)$

   (b) $(x^3 + 2x^2 - 3x - 4) \div (x + 2)$

   (c) $(x^3 - 4x^2 + 5x - 2) \div (x - 2)$

   (d) $(3x^3 + 4x^2 + x - 2) \div (x + 3)$

   (e) $(2x^3 - 3x^2 + 4x + 4) \div (2x + 1)$

2. Find the quotient and remainder in each of the following divisions.

   (a) $(x^3 + 2x^2 + 3x + 1) \div (x - 1)$

   (b) $(x^3 - 3x^2 + 4x - 2) \div (x - 2)$

   (c) $(x^3 + 5x^2 + 3x + 1) \div (x + 3)$

   (d) $(3x^3 - 2x^2 + 5x - 2) \div (x - 1)$

   (e) $(2x^3 - 7x^2 + 11x - 4) \div (2x - 1)$

   (f) $(2x^3 + 5x^2 + 6x - 8) \div (2x + 1)$

5.2 Factorizing Polynomials

Suppose that when $x^3 - 2x^2 - 5x + 8$ is divided by $x - 3$ the quotient is Q(x) and the remainder is R, then

$$x^3 - 2x^2 - 5x + 8 = (x - 3)Q(x) + R$$

If we substitute 3 for x we get

$$3^3 - 2 \cdot 3^2 - 5 \cdot 3 + 8 = 0 \cdot Q(3) + R$$

$$27 - 18 - 15 + 8 = R$$

$$2 = R$$

Therefore, the remainder when $x^3 - 2x^2 - 5x + 8$ is divided by $x - 3$ is 2.

You can verify this by using long division.

Notice that, the remainder is obtained by substituting into $x^3 - 2x^2 - 5x + 8$ the value of x that makes $(x - 3)$ zero.

The preceding discussion suggests the following theorem, called the Remainder Theorem.

The Remainder Theorem

The remainder when a polynomial P(x) is divided by $(x - a)$ is P(a).

A special case of the Remainder Theorem is the Factor Theorem.

The Factor Theorem

$x - \alpha$ is a factor of the polynomial $P(x)$ if and only if $P(\alpha) = 0$.

Examples

1. Find the remainder when $3x^3 + x^2 - 8x + 6$ is divided by $3x - 2$.

   Substitute 2/3 for x.

   The remainder is

   $$3\left(\frac{2}{3}\right)^3 + \left(\frac{2}{3}\right)^2 - 8\left(\frac{2}{3}\right) + 6 = \frac{8}{9} + \frac{4}{9} - \frac{16}{3} + 6 = 2$$

When finding factors of a polynomial, use try and error method to find one factor and then find a second factor by long division.

2. Factorize $2x^3 + 3x^2 - 11x - 6$ and find the zeros.

The first numbers to try are factors of 6. We will start with 2.

$$2(2^3) + 3(2^2) - 11(2) - 6 = 0$$

So $x - 2$ is a factor of the polynomial.

Dividing the polynomial by $x - 2$ gives $2x^2 + 7x + 3$.

Therefore

$$2x^3 + 3x^2 - 11x - 6 = (x - 2)(2x^2 + 7x + 3)$$

Complete the factorization by factorizing the quadratic expression. So,

$$2x^3 + 3x^2 - 11x - 6 = (x - 2)(x + 3)(2x + 1)$$

The zeros are -3, $-½$ and 2.

3. Find the value of a if $x - 2$ is a factor of $x^3 + ax^2 + x + 6$ and factories the expression completely.

Because $x - 2$ is a factor of $x^3 + ax^2 + x + 6$

$$2^3 + a(2^2) + 2 + 6 = 0$$

$$16 + 4a = 0$$

$$a = -4$$

Use long division to find another factor of $x^3 - 4x^2 + x + 6$. The result will be $x^2 - 2x - 3$. Hence,

$$x^3 - 4x^2 + x + 6 = (x - 2)(x^2 - 2x - 3)$$

$$= (x - 2)(x - 3)(x + 1)$$

Exercise 5.2

1. Find the remainder in each of the following divisions.

(a) $(x^3 + 3x^2 - 4x + 2) \div (x - 1)$

(b) $(x^3 + x^2 - 3x - 2) \div (x + 2)$

(c) $(x^3 + 3x^2 - 8x + 1) \div (x - 3)$

(d) $(4x^3 - 3x^2 + 2x - 7) \div (x + 1)$

(e) $(x^3 - 5x^2 + 11x - 6) \div (x - 2)$

(f) $(4x^3 - 6x^2 + 5) \div (2x - 1)$

2. Factorize the following expressions completely.

(a) $x^3 - 2x^2 - 5x + 6$

(b) $x^3 + 6x^2 + 11x + 6$

(c) $x^3 + 5x^2 - 2x - 24$

(d) $2x^3 + 3x^2 - 11x - 6$

3. Factorize completely each polynomial, given the indicated factors.

(a) $x^3 - 3x^2 - x + 3$; $x - 1$

(b) $x^3 + 2x^2 - 5x - 6$; $x - 2$

(c) $2x^3 - 3x^2 - 3x + 2$; $2x - 1$

(d) $2x^3 + x^2 - 18x - 9$; $2x + 1$

4. Find k, given that $x - 1$ is a factor of $3x^3 + kx^2 - 4x - 3$.

5. Find k, given that $x + 1$ is a factor of $3x^3 - kx + 7$.

6. Find p, given that $2x^3 + x^2 + px + 2p^2$ is divisible by $x + 1$.

7. Find p and q, given that $x^2 - 1$ is a factor of $x^4 + px^3 + qx - x + 2$.

8. $x - 2$ and $x + 1$ are both factors of $ax^3 + 3x^2 - 9x + b$. Find the value of a and of b. State the third factor of the expression.

9. When $x^3 - x^2 + ax + b$ is divided by $x - 1$ the remainder is -8. $x + 1$ is a factor of the expression. Find the value of a and of b and factorize the expression completely.

10. When $x^3 + ax^2 + bx - 2$ is divided by $x - 1$ the remainder is -4 and when it is divided by $x + 1$ the remainder is 4. Find the value of a and of b.

11. When the expression $x^3 + x^2 + px + q$ is divided by $x^2 - 1$, the remainder is $2x + 3$. Find the value of p and of q.

12. When $ax^4 + bx^3 + 3x^2 - 2x + 3$ is divided by $x^2 - 3x + 2$ the remainder is $x + 1$. Find the value of a and of b.

5.3 Solving Cubic Equations

You can solve the cubic equation $ax^3 + bx^2 + cx + d = 0$ by rewriting the equation in the form $(x + a)(pc^2 + qx + r)$, where a, p, q and r are integers.

Examples

1. Solve $3x^3 + 4x^2 - 17x - 6 = 0$

   Find a factor of the polynomial expression.

   Trying $x = 2$, we have

   $$3(2^3) + 4(2^2) - 17(2) - 6 = 0.$$

   So, $x - 2$ is a factor of the polynomial.

   Dividing $3x^3 + 4x^2 - 17x - 6$ by $x - 2$ gives $3x^2 + 10x + 3$. So,

   $$3x^3 + 4x^2 - 17x - 6 = (x - 2)(3x^2 + 10x + 3)$$

   Complete the factorization by factorizing the quadratic expression.

   Therefore $(x - 2)(3x + 1)(x + 3) = 0$

   Set each factor to zero, and then solve the linear equation.

   The solutions are $x = -3$, $x = -1/3$ and $x = 2$.

2. Solve $24 - 5x^2 = 2x - x^3$, given that $x + 2$ is a factor.

First, rewrite the equation in the general form.

$$x^3 - 5x^2 - 2x + 24 = 0$$

Dividing $x^3 - 5x^2 - 2x + 24$ by $x + 2$ gives $x^2 - 7x + 12$. So,

$$(x + 2)(x^2 - 7x + 12) = 0$$

$$(x + 2)(x - 3)(x - 4) = 0$$

The solutions of the equation are $x = -2$, $x = 3$ and $x = 4$.

Exercise 5.3

1. Solve the following equations.

    (a) $x^3 - 2x^2 - 5x + 6 = 0$ (b) $x^3 - 5x^2 - 2x + 24 = 0$

    (c) $x^3 + 3x^2 - 4x - 12 = 0$ (d) $x^3 - 4x^2 - 7x + 10 = 0$

    (e) $x^3 + 2x^2 - 11x - 12 = 0$ (f) $2x^3 + 3x^2 - 23x - 12 = 0$

2. Solve:

    (a) $x^3 + 2x^2 - 5x - 6 = 0$ given that $x + 1$ is a factor.

    (b) $x^3 + 6x^2 + 11x + 6 = 0$ given that $x + 2$ is a factor.

    (c) $2x^3 + 3x^2 - 3x - 2 = 0$ given that $x - 1$ is a factor.

    (d) $3x^3 + 4x^2 - 5x - 2 = 0$ given that $3x + 1$ is a factor.

Review Exercise 5

1. Divide:

    (a) $3x^3 + 4x^2 + 7x + 2$ by $3x + 1$ (b) $3x^3 + 2x^2 - 3x - 2$ by $3x + 2$

    (c) $2x^3 + 7x^2 - x + 12$ by $x + 4$ (d) $2x^3 - x^2 - 2x + 1$ by $2x - 1$

2. Find the remainder when:

    (a) $x^3 - 2x^2 + 5x + 8$ is divided by $x - 2$.

    (b) $12x^3 + 16x^2 - 5x - 3$ is divided by $2x - 1$.

(c) $2x^3 + x^2 - 13x + 6$ is divided by $x - 2$.

(d) $x^5 - 4x^3 + 2x + 3$ is divided by $x - 1$

3. Factorize completely:

   (a) $x^3 - 5x^2 + 2x + 8$ (b) $3x^3 - 8x^2 - x + 10$

   (c) $2x^3 - x^2 - 8x + 4$ (d) $x^3 - 7x + 6$

4. Show that $x - 3$ is a factor of $x^3 - 6x^2 + 11x - 6$. Find the other factors.

5. Factorize each polynomial and state its zeros.

   (a) $x^3 - 2x^2 - 5x + 6$ (b) $x^3 - 4x^2 + x + 6$

   (c) $x^3 - 6x^2 + 11x - 6$ (d) $2x^3 - x^2 - 8x + 4$

   (e) $2x^3 - x^2 - 8x + 4$ (f) $x^4 - 8x^3 + 14x^2 + 8x - 15$

6. Given that $3x^3 + x^2 + px + 12 = (x - 2)(ax^2 + bx + c)$ find the values of a, b, c and p. Hence, factorize completely $3x^3 + x^2 + px + 12$.

7. Given that $2x^3 - 3x^2 - 2x + p = (x + 1)(ax^2 + bx + c)$ find the values of a, b, c and p. Hence, factorize completely $2x^3 - 3x^2 - 2x + p$.

8. The polynomial $f(x) = 2x^3 - x^2 + px + 1$ has $(2x - 1)$ as a factor. Find the value of p and the zeros of $f(x)$.

9. $x^4 + 3x^3 + ax^2 + bx - 18$ is divisible by $x^2 - 9$. Find the value of a and of b.

10. The polynomial $f(x) = x^3 + ax^2 + bx - 12$ has $(x - 2)$ as a factor. When it is divided by $(x - 1)$ the remainder is -12. Find the remainder when $f(x)$ is divided by $(x - 3)$.

11. The polynomial $f(x) = x^3 - x^2 - 4kx + 3k^2$, where k is a constant, has $(x - 2)$ as a factor. Find the possible values of k.

12. Solve the following equations.

    (a) $3x^3 - 4x^2 - 12x + 16 = 0$ (b) $4x^3 - 7x - 3 = 0$

    (c) $4x^3 + 7x^2 - 5x - 6 = 0$ (d) $6x^3 - 5x^2 - 12x - 4 = 0$

6 Simultaneous Equations

Two linear equations with two unknowns which can be solved at the same time and have a common solution are called simultaneous equations.

6.1 Solving Simultaneous Linear Equations

Simultaneous linear equations can be solved algebraically by the method of elimination and the method of substitution.

Method of Elimination

This method allows us to eliminate one of the unknown by either adding or subtracting the two equations.

Examples

Solve the following simultaneous equations.

1.
$$3x + 2y = 13 \quad (1)$$
$$-3x + y = -7 \quad (2)$$

By adding equation (1) and equation (2) we eliminate the variable x.

$$3y = 6$$
$$y = 2$$

To find the value of x we substitute $y = 2$ into either equation (1) or equation (2).

Substituting $y = 2$ into equation (1) gives

$$3x + 4 = 13$$
$$3x = 9$$
$$x = 3$$

In this example it was possible to eliminate x by adding the equations because the coefficients of x are numerically equal and opposite in sign. Often neither x nor y can be eliminated by simply adding or subtracting the two equations. In such cases, multiply one or both of the equations by a number in order to make the coefficients of x (or y) numerically equal.

86 CHAPTER 6 Simultaneous Equations

$$3x + 4y = 17 \quad (1)$$

$$2x + 3y = 12 \quad (2)$$

$(1) \times 2 \quad 6x + 8y = 34 \quad (3)$

$(2) \times 3 \quad 6x + 9y = 36 \quad (4)$

$(3) - (4) \quad -y = -2$

$$y = 2$$

Substitute $y = 2$ into equation (2)

$$2x + 6 = 12$$

$$2x = 6$$

$$x = 3$$

Exercise 6.1(a)

Solve the following simultaneous equations.

1. $2x + y = 5$
   $x - y = 1$

2. $2x + y = 5$
   $2x + y = 8$

3. $x + 2y = 7$
   $3x - 2y = 9$

4. $3x + 2y = 8$
   $5x + 2y = -8$

5. $3x - 2y = 2$
   $3x + 4y = 14$

6. $3x + 2y = 4$
   $5x + 4y = 3$

7. $5x - 3y = -5$
   $3x + 11y = -3$

8. $2x - 5y = -14$
   $3x + 2y = 17$

9. $4x - 5y = 5$
   $2x - 3y = 2$

10. $-5x + 6y = -7$
    $4x + 3y = 16$

11. $5x + 2y = 2$
    $2x + 3y = -8$

12. $5x + 4y = 2$
    $-4x + 3y = 17$

Solving Simultaneous Linear Equations

Method of Substitution

To solve simultaneous equations by substitution, take one of the equations and express one of the unknowns in terms of the other. Then substitute this into the other equation to obtain an equation with one unknown.

Example

Solve the simultaneous equations.

$$3x - 2y = 8 \quad (1)$$
$$2x + y = 10 \quad (2)$$

Equation (2) can be rearranged to give

$$y = 10 - 2x$$

Substitute this into equation (1)

$$3x - 2(10 - 2x) = 8$$
$$3x - 20 + 4x = 8$$
$$7x = 28$$
$$x = 4$$

Substitute $x = 4$ into equation (2)

$$8 + y = 10$$
$$y = 2$$

Exercise 6.1(b)

Solve the following simultaneous equations.

1. $y = x + 1$
   $x + y = 3$

2. $y = 2x - 4$
   $3x + y = 11$

3. $x + y = 4$
   $2x - y = 5$

4. $y - 2x = 1$
   $3x - 4y = 1$

88 CHAPTER 6   Simultaneous Equations

5. $3x + 2y = 10$

   $4x - y = 6$

6. $4x + 3y = 9$

   $2x + 5y = 15$

7. $3x - 2y = 4$

   $2x + 3y = -6$

8. $5x + y = 1$

   $x + y = 5$

9. $x - y = 4$

   $-2x + 5y = -17$

10. $x + 2y = 4$

    $3x - 2y = -12$

11. $4x - 3y = 1$

    $x - 2y = 4$

12. $3x + 2y = 10$

    $4x - y = 6$

6.2 Simultaneous Equations with One Linear Equation

The method of substitution already explained can be used to solve simultaneous equations in two unknown with one linear equation.

Example

Solve the simultaneous equations

$$x^2 + 3xy + y^2 = -5 \qquad (1)$$

$$2x + y = 1 \qquad (2)$$

Begin by rearranging equation (2)

$$y = 1 - 2x$$

Substitute this into equation (1)

$$x^2 + 3x(1 - 2x) + (1 - 2x)^2 = -5$$

Simplifying this equation we get

$$x^2 + x - 6 = 0$$

$$(x - 2)(x + 3) = 0$$

$$x - 2 = 0 \quad \text{or} \quad x + 3 = 0$$

$$x = 2 \qquad x = -3$$

Substitute $x = 2$ into (2)

$$4 + y = 1$$
$$y = -3$$

Again substitute $x = -3$ into (2)

$$-6 + y = 1$$
$$y = 7$$

The solutions are $x = 2$, $y = -3$ and $x = -3$, $y = 7$.

Exercise 6.2

Solve the simultaneous equations.

1. $x^2 + 2xy + y^2 = 9$
   $2x + y = 5$

2. $2x^2 + y^2 = 19$
   $3x + y = 10$

3. $2x - y = 2$
   $y^2 - 2x^2 - 14 = 0$

4. $3y - x = -1$
   $x^2 - 2xy - y^2 = 7$

5. $2x - y = 1$
   $3x^2 - xy + y^2 = 15$

6. $x^2 + 2xy - 4y^2 = -4$
   $3x - 2y = 8$

7. $2x + y = 3$
   $2x^2 + 3xy + 2y^2 = 9$

8. $3x - 2y = 1$
   $9x^2 - 4y^2 = 17$

9. $x^2 + 2xy = 8$
   $x + 2y = 2$

10. $x^2 + y^2 + 2xy + 6x = -2$
    $2x - y = -5$

Review Exercise 6

Solve the following simultaneous equations.

1. $2x + y = 3$
   $2x + y = 9$

2. $3x + 2y = 4$
   $5x - 2y = 12$

3. $5a - 3b = 11$
   $a + b = 7$

4. $3x + 2y = 4$
   $5x + y = 9$

5. $4a - 3b = 5$
   $7a - 6b = 5$

6. $7x + 5y = 3$
   $3x + 2y = 1$

7. $2x + 3y = 13$
   $7x - 5y = -1$

8. $6x - 7y = -1$
   $5x - 4y = 12$

9. $3x + 2y = 20$
   $-5x + 3y = 11$

10. $2x + 3y = 17$
    $-3x + y = -9$

11. $3x - 2y = 1$
    $5x - 3y = 3$

12. $4x - 3y = 1$
    $x - 2y = 4$

13. $\dfrac{x}{6} + \dfrac{y}{3} = 8$
    $\dfrac{x}{4} - \dfrac{y}{9} = 1$

14. $\dfrac{a}{2} - \dfrac{b}{3} = 1$
    $\dfrac{a}{4} - \dfrac{b}{9} = \dfrac{2}{3}$

15. $x^2 + y^2 = 10$
    $y - x = 4$

16. $x^2 + y^2 - 4 = 0$
    $x + y = 2$

17. $2x^2 + y^2 = 18$
    $y = x - 3$

18. $2x^2 + xy + y^2 = 2$
    $x + y = 1$

19. $x^2 - 2xy + y^2 = 36$
    $X + y = 2$

20. $2y = x + 3$
    $x^2 + y^2 - 2x + 6y = 15$

7 Exponential and Logarithmic Functions

7.1 Exponential Functions

An exponential function is defined by an equation of the form $y = a^x$, where $a > 0$, $a \neq 1$ and x is a real number. Examples include $y = 2^x$, $y = x^x$ and $y = 10^x$.

The table shows some values of $y = 2^x$..

x	−2	−1	0	1	2	3
y	¼	½	1	2	4	8

The graph of the function is shown in Figure 7.1.

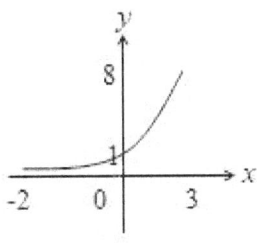

Figure 7.1

Notice that the value of the function increases as x increases.

Now, we consider the function $y = 2^{-x}$.

The table shows some values of $y = 2^{-x}$.

x	−3	−2	−1	0	1	2
y	8	4	2	1	½	¼

The graph of the function is shown in Figure 7.2.

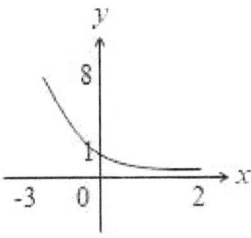

Figure 7.2

We notice that in this case the values of the function decrease as x increases. You see that the point $(0, 1)$ is the y-intercept for each of the graphs, and the graphs do not cross the x-axis.

In general, an exponential function has the following properties.

1. The values of the function $y = a^x$ increase as x increases.

2. The values of the function $y = a^{-x}$ decrease as x increases.

3. The graph has a y-intercept at $(0, 1)$.

4. The function is one-to-one.

5. The domain is the set of all real numbers

6. The range is the set of all positive numbers.

The Natural Exponential Functions

The function $y = e^x$ is called the natural exponential function. e is called the Euler's number and is approximately 2.7182818. The natural exponential function has the same properties as listed for the exponential function $y = a^x$.

The rules of indices can be extended to cover exponential functions.

Sketching the Graphs of Exponential Functions

The following transformations would help you use the graph of $y = e^x$ to sketch graphs of functions of the form $y = ke^x + a$.

1. Translation in the y-direction.

    $f(x) + a$ moves the graph of $f(x)$ a unit up (down if a is negative).

2. Translation in the x-axis.

    $f(x + a)$ moves the graph of $f(x)$ a unit to the left (right if a is negative).

3. Reflection in the x-axis.

    The graph of $y = -f(x)$ is the graph of $y = f(x)$ reflected in the x-axis.

4. Reflection in the y-axis.

    The graph of $y = f(-x)$ is the graph of $y = f(x)$ reflected in the y-axis.

Exponential Functions 93

5. The graph of $y = af(x)$ is obtained by stretching the graph of $y = f(x)$ along the y-axis by a factor of a.

Examples

Sketch the graph of each of the following functions.

1. $g(x) = 2^{x+1}$

We can use the graph of $f(x) = 2^x$ to sketch the graph of g.

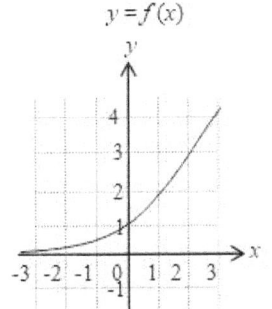
Figure 7.3

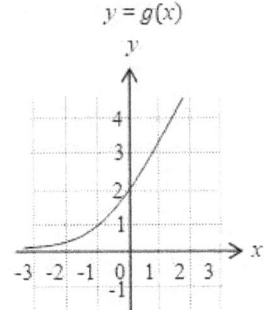
Figure 7.4

The function g is related to f by $g(x) = f(x + 1)$. To sketch the graph of g, shift the graph of f, shown in Figure 7.3, one unit to the left as shown in Figure 7.4. Note that, the y-intercept of g is $(0, 2)$.

2. $h(x) = 2^x - 3$

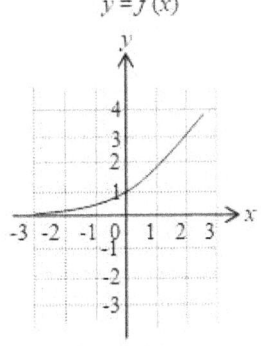
Figure 7.5

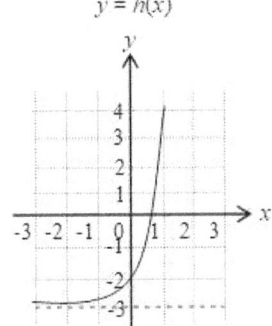
Figure 7.6

The function h is related to f by $h(x) = f(x) - 3$. To sketch the graph of h, shift the graph of f three units downward, as shown in Figure 7.6.

7.2 Logarithmic Functions

Recall that $y = 2^x$ is one-to-one function and so has an inverse function. The inverse function is called the logarithmic function. You can obtain the inverse function by interchanging the roles of x and y. That is, the inverse function is $x = 2^y$. The table shows some values of $x = 2^y$.

x	¼	½	0	1	2	3
$x = 2^y$	−2	−1	0	1	2	3

The graph of the function is shown in Figure 7.7.

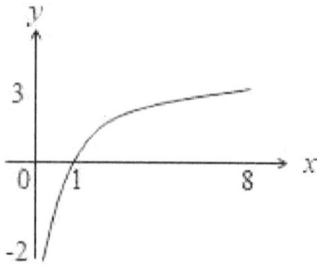

Figure 7.7

The following properties of the logarithmic functions hold.

1. The function is one-to-one.

2. The domain is the set of all positive real numbers.

3. The range is the set of all real numbers.

4. The graph intersects the x-axis at (1, 0).

Definition of Logarithmic Functions

The logarithm of x with base a, denoted by $\log_a x$, is the power to which a must be raised in order to get x, i.e. $x = a^y$.

For example, $8 = 2^3$, so $\log_2 8 = 3$.

The Natural Logarithmic Functions

As with exponential functions, we sometimes use e as a base for a logarithmic function. The logarithmic function with base e is called the natural logarithmic function and is denoted by $\ln x$.

Common Logarithms

The logarithmic function with base 10 is called the common logarithmic function. It is customary to omit the base in writing a common logarithm. When no base is shown we take the base as 10.

Sketching the Graphs of Logarithmic Functions

Examples

Sketch the graph of each of the following functions.

1. $g(x) = 2 + \ln x$

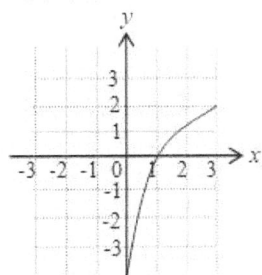
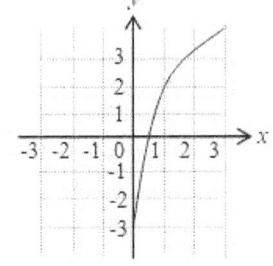

Figure 7.8   Figure 7.9

Because $g(x) = 2 + \ln x = 2 + f(x)$, the graph of g can be obtained by shifting the graph of f two units upward as shown in Figure 7.9.

2. $h(x) = \ln(x + 3)$

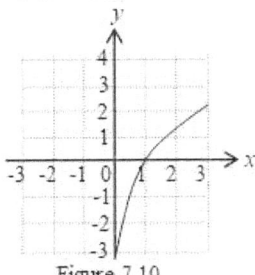
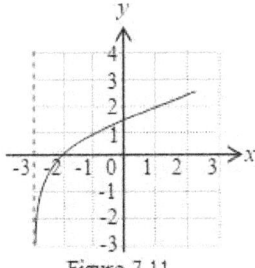

Figure 7.10   Figure 7.11

Because $h(x) = \ln(x + 3) = f(x + 3)$, the graph of h can be obtained by shifting the graph of f three units to the left.

CHAPTER 7 Exponential and Logarithmic Equations

Evaluating Logarithms

Some logarithms can be calculated by writing the logarithm in index form.

Examples

Evaluate:

1. $\log_2 32$

$$\text{Let } x = \log_2 32$$
$$2^x = 32$$
$$2^x = 2^5$$
$$x = 5$$

So, $\log_2 32 = 5$

2. $\log_3 \dfrac{1}{81}$

$$\text{Let } x = \log_3 \dfrac{1}{81}$$
$$3^x = \dfrac{1}{81}$$
$$3^x = 3^{-4}$$
$$x = -4$$

So, $\log_3 \dfrac{1}{81} = -4$

Example 7(a)

Evaluate:

1. $\log_3 243$
2. $\log_4 64$
3. $\log_2 \dfrac{1}{128}$
4. $\log 1000$
5. $\log_6 \dfrac{1}{36}$
6. $\log_{25} 5$
7. $\log_{\sqrt{2}} 8$
8. $\log_{0.1} 100$

Properties of Logarithms

The following properties of logarithms follow directly from the definition of logarithmic function with base a.

For any positive real number a such that $a \neq 1$, $a^0 = 1$ and $a^1 = a$, then

1. $\log_a 1 = 0$

2. $\log_a a = 1$

Further properties of logarithms can be derived using the rules of indices.

Product Rule

Let $\log_a x = p$ and $\log_a y = q$

Then $x = a^p$ and $y = a^q$

So, $x \cdot y = a^p \cdot a^q = a^{p+q}$

Writing this in logarithmic form we get

$$\log_a xy = p + q$$

Substituting $\log_a x$ and $\log_a y$ for p and q respectively gives

$$\log_a xy = \log_a x + \log_a y$$

The following properties of logarithm hold. Some of the properties have not been proved. We leave the proof as an exercise.

1. $\log_a xy = \log_a x + \log_a y$

2. $\log_a \dfrac{x}{y} = \log_a x - \log_a y$

3. $\log_a x^n = n \log_a x$

4. $\log_a a = 1$

5. $\log_a 1 = 0$

Expanding Logarithmic Expressions

We can use the properties of logarithms to expand logarithmic expressions.

Examples

Expand each of the following expressions.

1. $\log_a \dfrac{a^3 b^4}{c^2}$

$$\log_a \dfrac{a^3 b^4}{c^2} = \log_a a^3 b^4 - \log_a c^2$$
$$= \log_a a^3 + \log_a b^4 - \log_a c^2$$
$$= 3\log_a a + 4\log_a b - 2\log_a c$$
$$= 3 + 4\log_a b - 2\log_a c$$

2. $\log \sqrt[4]{\dfrac{x^2}{y^6 z^4}}$

$$\log \sqrt[4]{\dfrac{x^2}{y^6 z^4}} = \dfrac{1}{4}(\log x^2 - \log y^6 z^4)$$
$$= \dfrac{1}{4}(2\log x - 6\log y - 4\log z)$$
$$= \dfrac{1}{2}\log x - \dfrac{3}{2}\log y - \log z$$

Exercise 7(b)

Expand each of the following expressions.

1. $\log_a x^3 y^2$
2. $\log_a \dfrac{x^2}{y}$
3. $\log x^2 \sqrt[3]{y}$
4. $\log_2 \dfrac{1}{x^3}$
5. $\ln x^3 \sqrt{y}$
6. $\log_3 \dfrac{x^2 y^3}{z^2}$
7. $\log_5 \dfrac{x^2 y}{\sqrt[3]{z}}$
8. $\ln \dfrac{x^4 \sqrt[3]{y}}{z^3}$
9. $\log \sqrt[4]{\dfrac{xy^2}{z^2}}$
10. $\ln \sqrt[3]{\dfrac{x^3 y^6}{z^9}}$
11. $\log_2 \dfrac{x^2 y^3}{\sqrt{z^5}}$
12. $\log_5 \sqrt[3]{\dfrac{1000 x^2}{y^3 z^4}}$

Writing Logarithmic Expressions as Single Logarithms

Examples

Write each of the following expressions as single logarithm.

1. $2\ln x + 3\ln y - 5\ln z$

$$2\ln x + 3\ln y - 5\ln z = \ln x^2 + \ln y^3 - \ln z^5$$
$$= \ln x^2 y^3 - \ln z^5$$
$$= \ln \frac{x^2 y^3}{z^5}$$

2. $5\log a + 3 - \frac{1}{2}\log b$

$$5\log a + 3 - \frac{1}{2}\log b = \log a^5 + \log 1000 - \log b^{1/2}$$
$$= \log 1000 a^5 - \log \sqrt{b}$$
$$= \log \frac{1000 a^5}{\sqrt{b}}$$

Exercise 7(c)

Express as single logarithms.

1. $\log_2 x^2 + \log_2 y^3$

2. $2\log_3 x - 3\log_3 y - \log_3 z$

3. $\log_3 x - 2\log_3 y$

4. $\frac{1}{2}\log_5 x - \frac{2}{3}\log_5 y - 3\log_5 z$

5. $3\log x - 9\log y + 6\log z$

6. $3\log x - \log y$

7. $\frac{2}{3}\log x + 2\log y$

8. $\frac{1}{2}\ln x + \frac{3}{4}\ln y - \ln z$

9. $\log_5 x + \frac{2}{3}\log_5 y$

10. $3\ln x + 2\ln y - \ln z$

11. $3 + 2\log x - \frac{1}{4}\log y$

12. $\frac{2}{3}\log x - 3\log y + 1$

Simplifying Logarithmic Expressions

Example

Simplify:

1. $\log_2 64$

$$\log_2 64 = \log_2 2^6$$
$$= 6 \log_2 2$$
$$= 6$$

2. $\log_3 \sqrt[4]{27}$

$$\log_3 \sqrt[4]{27} = \log_3 27^{1/4}$$
$$= \log_3 3^{3/4}$$
$$= \frac{3}{4} \log_3 3$$
$$= \frac{3}{4}$$

3. $\dfrac{\ln 27}{\ln 9}$

$$\frac{\ln 27}{\ln 9} = \frac{\ln 3^3}{\ln 3^2}$$
$$= \frac{3 \ln 3}{2 \ln 3}$$
$$= \frac{3}{2}$$

Exercise 7(d)

Simplify:

1. $\log 100000$
2. $\log_2 128$
3. $\log_3 27$
4. $\log_6 216$
5. $\log_4 \frac{1}{16}$
6. $\log_5 125$
7. $\frac{4}{5} \log_2 32$
8. $\log_{\frac{1}{3}} 27$
9. $\log_8 \frac{1}{2}$

10. $\log_2 \sqrt[3]{2}$ 11. $\dfrac{\log_2 81}{\log_2 27}$ 12. $\dfrac{\ln 64}{\ln 8}$

Change of Base

Most calculators can be used to evaluate logarithms with base 10 and base e. You can use a calculator that has either the common logarithm key or the natural logarithm key to evaluate logarithm to any base by using change of base formula.

Suppose that

$$b^x = a$$

where a, b and x are positive real numbers such that $a \neq 1$ and $b \neq 1$.

Taking logarithm to base c of both sides gives

$$\log_c b^x = \log_c a$$

$$x \log_c b = \log_c a$$

$$x = \frac{\log_c a}{\log_c b}$$

Because $x = \log_b a$ we have

$$\log_b a = \frac{\log_c a}{\log_c b}$$

This equation is the change of base formula.

Examples

Evaluate:

1. $\log_2 25$

$$\log_2 25 = \frac{\log 25}{\log 2}$$

$$= \frac{1.3979}{0.3010}$$

$$= 4.644$$

2. $\log_5 16$

$$\log_5 16 = \frac{\log 16}{\log 5}$$

$$= \frac{2.7726}{1.6094}$$

$$= 1.732$$

You will get the same answer when you use natural logarithms in the change of base formula.

Exercise 7(e)

Evaluate:

1. $\log_2 26$
2. $\log_4 9$
3. $\log_5 10$
4. $\log_3 8$
5. $\log_7 12$
6. $\log_6 15$
7. $\log_8 32$
8. $\log_9 8$
9. $\log_6 10$
10. $\log_5 16$
11. $\log_4 13$
12. $\log_{12} 64$

Solving Exponential Equations

In Chapter 4, you were able to solve exponential equation because each side of the equation could be written in index form with the same base. For other equations such as $3^x = 25$, it will be difficult to write the expression on both sides in index form with the same base. In cases like this we rewrite the equation in logarithmic form by taking the logarithm of each side.

Examples

Solve each of the following equations.

1. $3^x = 7$

$$\ln 3^x = \ln 7$$

$$x \ln 3 = \ln 7$$

$$x = \frac{\ln 7}{\ln 3} = 1.771$$

2. $\left(\frac{1}{4}\right)^{x-2} = 5^{3x}$

$$\log\left(\frac{1}{4}\right)^{x-2} = \log 5^{3x}$$

$$\log(2^{-2})^{x-2} = \log 5^{3x}$$

$$(-2x + 4)\log 2 = 3x \log 5$$

$$x = \frac{4 \log 2}{2 \log 2 + 3 \log 5}$$

$$= 0.446$$

Exercise 7(f)

Solve:

1. $5^x = 8$
2. $4^x = 20$
3. $3^{2x} = 4$
4. $7^{3x} = 50$
5. $2^{2x+1} = 15$
6. $4^x = 3^{x+1}$
7. $6^{x-2} = 9$
8. $3^{x+4} = 6$
9. $e^{2-x} + 7 = 21$
10. $15 - 8e^x = 6$
11. $5^{x+6} - 7 = 18$
12. $e^{x+1} = 15$

Solving Logarithmic Equations

A logarithmic equation is an equation that contains a logarithmic expression. To solve a logarithmic equation, first isolate the logarithmic expression then write the equation in index form.

Examples

Solve:

1. $\log_3 x + \log_3 2 = 4$

$$\log_3 2x = 4$$

$$2x = 3^4$$

$$2x = 81$$

$$x = \frac{81}{2}$$

$$= 40\frac{1}{2}$$

You need to check the answer in the original equation to avoid answers that would result in taking logarithm of zero or negative numbers.

$$\log_3\left(\frac{81}{2}\right) + \log_3 2 = \log_3 81 - \log_3 2 + \log_3 2$$

$$= 4\log_3 3$$

$$= 4$$

So the solution is $x = 40\ \frac{1}{2}$

2. $\log_2 \frac{1}{2}x + \log_2(x-4) = 4$

$$\log_2 \frac{1}{2}x(x-4) = 4$$

$$\frac{1}{2}x(x-4) = 2^4$$

$$x(x-4) = 32$$

$$x^2 - 4x - 32 = 0$$

$$(x-8)(x+4) = 0$$

$$x = -4 \text{ or } x = 8$$

Substituting -4 for x in the original equation gives

$$\log_2(-2) + \log_2(-8) \neq 4$$

Substituting 8 for x in the original equation gives

$$\log_2 4 + \log_2 4 = 2\log_2 4 = 4$$

The only solution is $x = 8$

3. $\log(2x + 6) - \log(x + 3) = \log x$

$$\log(2x + 6) - \log(x + 3) - \log x = 0$$

$$\log \frac{2x+6}{x(x+3)} = 0$$

$$\frac{2x+6}{x^2+3x} = 10^0$$

$$2x + 6 = x^2 + 3x$$

$$x^2 + x - 6 = 0$$

$$(x - 2)(x + 3) = 0$$

$$x = -3 \text{ or } x = 2$$

The only solution is $x = 2$. Check the answers in the original equation.

You can obtain the same answer as follows.

$$\log(2x + 6) - \log(x + 3) = \log x$$

$$\log \frac{2x+6}{x+3} = \log x$$

Therefore, $\quad \dfrac{2x+6}{x+3} = x$

$$x^2 + x - 6 = 0$$

$$(x - 2)(x + 3) = 0$$

$$x = -3 \text{ or } x = 2$$

The solution is $x = 2$

Exercise 7(g)

Solve:

1. $\log_2(x + 1) = 3$
2. $\log(2x + 1) - \log(x + 2) = \log x$
3. $\log_2 x + \log_2 8 = 6$
4. $\log_3 x + \log_3(2x + 3) = 2$
5. $\log_5(2x - 1) = 2.$
6. $\log_2(x + 2) + \log_2(x - 5) = 3$

7. $\log x + \log 5 = 2$
8. $\log_3(x+1) - \log_3(x-2) = 2$
9. $\log(x+2) - \log(2x-1) = 1$
10. $\log_x 7 + 1 = \log_x 112 - 3$
11. $\log(x+9) - \log(x-2) = \log 5$
12. $\log_3 x - 2\log_x 3 = 1$

Review Exercise 7

1. Use the properties of logarithms to expand the following expressions.

   (a) $\log\left(\frac{x^2 y}{z^3}\right)$

   (b) $\ln\left(\frac{x^5 y^3}{z^2}\right)$

   (c) $\log_2\left(\frac{16x^2 y^3}{z^4}\right)$

   (d) $\frac{1}{2}\ln\left(\frac{x^4 y^6}{z^8}\right)$

   (e) $\log_5 \sqrt{\frac{x^5}{y^8 z^7}}$

   (f) $\log \sqrt[3]{\frac{x^6 y^2}{z^4}}$

2. Use the properties of logarithms to write the following expression as single logarithms.

   (a) $3\log x + 5\log y$

   (b) $2\log x - 3\log y$

   (c) $5\ln x + 3\ln y - 2\ln z$

   (d) $\frac{3}{5}\log_2 x - \frac{2}{3}\log_2 y$

   (e) $\frac{1}{2}\log x + \frac{3}{4}\log y - 3\log z$

   (f) $2\ln x + \frac{1}{4}\ln y$

   (g) $\frac{1}{3}\ln x - 3\ln y + 2\ln z$

   (h) $3\log_2 x - 8 - 2\log_2 y$

3. Solve:

   (a) $3^x = 91$

   (b) $5^x = 8$

   (c) $7^{3-x} = 15$

   (d) $9^{3x} = 126$

4. Solve:

   (a) $\log_6(x-5) + \log_6 x = 2$

   (b) $2\log_4(x+5) = 3$

   (c) $\log x + \log(x-3) = 1$

   (d) $\log_3 x - \log_3 4 = 2$

   (e) $\ln(x+3) + \ln x = \ln 10$

   (f) $2\log_4 x - \log_4(x-1) = 1$

8 Straight Line Graphs

8.1 The Distance between Two Points

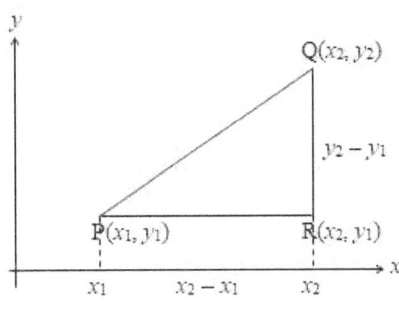

Figure 8.1

Figure 8.1 shows a straight line joining the points P and Q having coordinates (x_1, y_1) and (x_2, y_2) respectively. PR and QR are drawn parallel to the x-axis and the y-axis respectively to form triangle PQR. By the Pythagoras' theorem

$$PQ^2 = PR^2 + QR^2$$
$$= (x_2 - x_1)^2 + (y_2 - y_1)^2$$
$$PQ = \sqrt{(x_2 - x_1)^2 + (y_2 - y_1)^2}$$

The distance between the points (x_1, y_1) and (x_2, y_2) is given by $\sqrt{(x_2 - x_1)^2 + (y_2 - y_1)^2}$.

Example

Find the distance between the points (2, 3) and (5, 7).

Let $(x_1, y_1) = (2, 3)$ and $(x_2, y_2) = (5, 7)$. Then

$$\text{Distance} = \sqrt{(5 - 2)^2 + (7 - 3)^2}$$
$$= \sqrt{9 + 16}$$
$$= \sqrt{25}$$
$$= 5 \text{ units}$$

CHAPTER 8 Straight Line Graphs

Exercise 8.1

1. Find the distance between the following pairs of points.

    (a) (3, 2), (8, 14)
    (b) (1, 3), (4, 7)
    (c) (−1, −1), (−1, 7)
    (d) (2, 4), (5, 2)
    (e) (4, −1), (6, 2)
    (f) (−4, −1), (−2, −3)

2. A triangle has vertices A (1, 1), B (4, 4) and C (9, −1). Calculate the lengths of the sides of the triangle and prove that the triangle is right-angled.

3. Show that the triangle whose vertices are (3, −2), (−1, 1) and (2, 5) is isosceles.

4. The point (5, t) is equidistant from (2, −1) and (1, 6). Find the value of t.

5. The point (t, − 1) lies on a circle with a radius of 5 and its centre at (−1, 3). Find the possible values of t.

6. Three points A, B and C have coordinates (9, 6), (3, −2) and (−5, h). Given that h is positive and AB = BC, find the value of h.

8.2 The Midpoint of a Straight Line

The midpoint of the straight line segment joining the points (x_1, y_1) and (x_2, y_2) is given by

$$\text{Midpoint} = \left(\frac{x_1+x_2}{2}, \frac{y_1+y_2}{2}\right)$$

Example

Find the midpoint of the line segment joining the points (8, 3) and (−2, 5)

$$\text{Midpoint} = \left(\frac{8+(-2)}{2}, \frac{3+5}{2}\right) = (3, 4)$$

Exercise 8.2

1. Find the coordinates of the midpoint of the line segments joining the following pairs of points.

   (a) (1, 2), (5, 4)   (b) (4, 2), (6, 10)

   (c) (−2, 3), (8, −4)   (d) (6, −5), (2, 3)

   (e) (−1, −6), (−4, 3)   (f) (7, −4), (−3, 8)

2. (a) Find the coordinates of the midpoints of the sides of the triangle whose vertices are A (3, 2), B (7, 4) and C (−6, −5).

   (b) Show that the length of the segment joining the midpoint of two sides of triangle ABC in (a) is one-half of the length of the third side.

3. The midpoint of a line segment AB is (3, 2). Given that the coordinates of A is (2, −4), find the coordinates of B.

4. If (6, 2) is the midpoint of the line segment connecting (3, −1) to P(x, y), find the value of x and of y.

The Gradient of a Straight Line

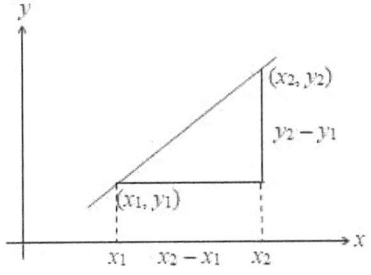

Figure 8.2

Figure 8.2 shows a straight line passing through the points (x_1, y_1) and (x_2, y_2). The gradient m of the line is the ratio of its y-change to its x-change.

$$\text{Gradient} = \frac{\text{Change in y}}{\text{Change in x}}$$

$$= \frac{y_2 - y_1}{x_2 - x_1}$$

CHAPTER 8 Straight Line Graphs

Example

Find the gradient of the straight line joining the points (3, 2) and (5, 6).

Let $(x_1, y_1) = (3, 2)$ and $(x_2, y_2) = (5, 6)$

$$\text{Gradient} = \frac{6-2}{5-3}$$

$$= 2$$

Exercise 8.3

1. Find the gradients of the lines joining the following pairs of points.

   (a) (5, 4), (9, 11)
   (b) (−4, 9), (2, −3)
   (c) (4, −3), (−7, −4)
   (d) (−1, −2), (5, 7)
   (e) (6, 7), (11, 3)
   (f) (−3, 5), (9, −5)

2. Prove, using gradients that the points (−6, 6), (−2, 4) and (6, 0) are collinear (that is the points lie on the same straight line).

3. Determine whether the sets of points are collinear.

   (a) (−1, −2), (4, 3), (5, 4)
   (b) (−4, 2), (−6, 5), (5, 10)
   (c) (3, 5), (−2, −5), (−3, 9)
   (d) (0, 6), (3, −3), (−1, 9)

4. If the points (−4, 3), (−1, 4) and (x, 9) are collinear then find x.

5. The points (−1, −1), (3, 11) and (1, t) lie on the same line. What is the value of t?

6. The points A(3, x) and B(7, 5) lie on a straight line. If the gradient of the line is ¾, find the value of x.

7. The points (3, 2) and (x, 5) lie on the same line. If the gradient of the line is − 3/2, find the value of x.

8. Three points A(a, b), B(4/3, a) and C(b, 5/2) lie on a straight line. If the gradient of the line is − 3/2, find the value of a and of b.

8.4 The Equation of a Straight Line

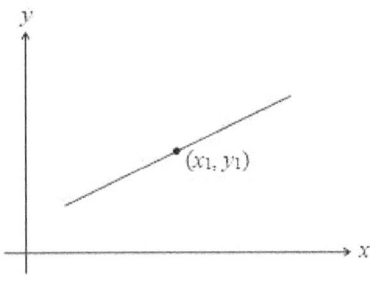

Figure 8.3

Figure 8.3 shows a straight line which passes through the point (x_1, y_1). If (x, y) is any other point on the line, then

$$\frac{y-y_1}{x-x_1} = m$$

$$y - y_1 = m(x - x_1)$$

which is called the point-gradient form of the equation of a line.

Other forms of the equation of a straight line are $y = mx + c$ and $ax + by + c = 0$.

The equation of a straight line in the form $y = mx + c$ is called the gradient-intercept form. The value m represents the gradient of the straight line, and c the intercept with the y-axis. When m is positive the graph of $y = mx + c$ is increasing and when m is negative the graph is decreasing as shown in Figure 8.4.

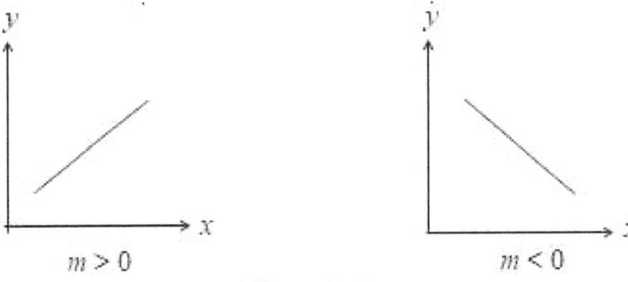

Figure 8.4

CHAPTER 8 Straight Line Graphs

Examples

1. Find the equation of the straight line which passes through $(2, -1)$ and $(1, -4)$.

   Begin by finding the gradient m.

   $$m = \frac{-4+1}{1-2} = 3$$

   Using the point $(2, -1)$ and the point-gradient form of equation of line, gives

   $$y - y_1 = m(x - x_1)$$

   $$y + 1 = 3(x - 2)$$

   $$y = 3x - 7$$

2. Find the equation of the straight line which passes through $(-1, 2)$ and $(3, 5)$.

   $$m = \frac{5-2}{3+1} = \frac{3}{4}$$

   The equation of the straight line is

   $$y - 2 = \frac{3}{4}(x + 1)$$

   $$3x - 4y + 11 = 0$$

Exercise 8.4

Find the equation of the straight line which passes through each of the following points.

1. $(2, 2), (5, 8)$
2. $(2, 5), (4, 11)$
3. $(2, -3), (3, 1)$
4. $(4, 1), (6, -5)$
5. $(-4, 2), (-2, 8)$
6. $(-2, 2), (-4, 5)$
7. $(-3, 6), (3, -2)$
8. $(2, -7), (6, -10)$
9. $(4, 3), (8, -2)$
10. $(4, 3), (8, 6)$
11. $(-1, -6), (2, -4)$
12. $(-3, -2), (-5, -5)$

8.5 Parallel and Perpendicular Lines

Recall from geometry that two lines in a plane are parallel if they do not intersect and that two lines are perpendicular if they intersect at right angles.

Parallel Lines

Two lines are parallel if and only if the gradients of the lines are equal. We will use this fact to write equations of straight line parallel to given lines.

Example

Find the equation of the straight line through $(3, -2)$ parallel to the line $4x + 3y - 12 = 0$.

First write $4x + 3y - 12 = 0$ in the gradient-intersect form.

$$y = -\frac{4}{3}x + 4$$

Because the line through $(3, -2)$ is parallel to the given line it has a gradient of $-4/3$.

Substituting $x_1 = 3$, $y_1 = -2$ and $m = -4/3$ into $y - y_1 = m(x - x_1)$ gives

$$y + 2 = -\frac{4}{3}(x - 3)$$

$$3y + 6 = -4x + 12$$

$$4x + 3y - 6 = 0$$

Perpendicular Lines

Two lines are perpendicular if and only if the product of their gradients is -1. That is, if two perpendicular lines have gradients m_1 and m_2, then $m_1 \times m_2 = -1$. It follows that two lines are perpendicular if their gradients are negative reciprocal of each other.

Example

Find the equation of the line through $(2, 3)$ and perpendicular to the line $3x - 2y - 9 = 0$.

Begin by writing $3x - 2y - 9 = 0$ in gradient-intersect form.

$$y = \frac{3}{2}x - \frac{9}{2}$$

Because the line through (2, 3) is perpendicular to the given line its gradient is $-2/3$.

Substituting $x_1 = 2$, $y_1 = 3$ and $m = -2/3$ into $y - y_1 = m(x - x_1)$ gives

$$y - 3 = -\frac{2}{3}(x - 2)$$

$$3y - 9 = -2x + 4$$

$$2x + 3y - 13 = 0$$

Exercise 8.5

1. Find the equation of the line through each point and parallel to the given line.

   (a) (4, 3), $4x - 3y + 12 = 0$
   (b) (−1, 4), $5x - 2y - 1 = 0$
   (c) (−3, 2), $2x + 3y - 5 = 0$
   (d) (−2, −1), $3x - 2y + 7 = 0$

2. Find the equation of the line through each point and perpendicular to the given line.

   (a) (4, 3), $5x + 3y - 10 = 0$
   (b) (−2, 5), $x + 3y + 2 = 0$
   (c) (−3, 2), $3x + 2y - 7 = 0$
   (d) (3, −4), $2x - 3y + 8 = 0$
   (e) (1, 2), $2y + 3x - 1 = 0$
   (f) (2, 3), $2x - 3y + 5 = 0$

3. A line passing through (−2, 1) and (3, y) is parallel to a line with gradient 2. Find the value of y.

4. A line passing through (5, 5) and (x, 8) is parallel to the line $3x + 2y + 8 = 0$, find the value of x.

5. Find the equation of the line which passes through the midpoint of the line joining the points (2, 1) and (5, 3) and parallel to the line $3x + 2y - 8 = 0$.

6. Find the equation of the perpendicular bisector of the line joining the points (2, 1) and (6, 5).

7. A line passing through (3, 2) and (y, 5) is perpendicular to a line with gradient 4/3. Find the value of y.

8. A line passing through (x, −2) and (3, −1) is perpendicular to the line $3x + 4y - 8 = 0$. Find the value of x.

9. Find the equation of the line through the midpoint of the line joining the points A (−4, −1) and B (8, 7) and perpendicular to AB.

10. Three points A (1, 1), B (4, 3) and C (−1, 4) are the vertices of a triangle. Show that ABC is a right-angled triangle. What is its area?

8.6 Point of Intersection

Two lines are said to intersect if and only if they have exactly one point in common. The common point is called the point of intersection. This point satisfies the equation of the lines. It follows that to find the point of intersection of two lines you need to solve simultaneous equations.

Example

Find the point of intersection of the lines $3x + 2y = 4$ and $2x - 3y = 7$.

We solve two simultaneous equations as follows.

$$3x + 2y = 4 \quad (1)$$

$$2x - 3y = 7 \quad (2)$$

(1) × 3 $9x + 6y = 12$ (3)

(2) × 2 $4x - 6y = 14$ (4)

(3) + (4) $13x = 26$

$x = 2$

Substitute $x = 2$ into equation (1)

$$3(2) + 2y = 4$$

$$y = -1$$

The point of intersection is (2, −1).

Exercise 8.6

1. Find the point intersection of each of the following pairs of lines.

    (a) $3x + y = 11$, $2x + y = 8$

    (b) $2x - 5y + 9 = 0$, $3x + 2y - 15 = 0$

    (c) $3x + 2y - 4 = 0$, $5x + 4y - 3 = 0$

    (d) $5x + 2y - 2 = 0$, $2x + 3y + 8 = 0$

    (e) $3x + 2y - 10 = 0$, $4x - y - 6 = 0$

2. The perpendicular bisector of the line joining the points (−3, 7) and (5, 3) meets the line $2y = x + 15$ at the point N. Find the coordinates of N.

3. The line from the point (1, 2) parallel to the line $3x + 4y - 8 = 0$ meets the line $3y = 4x + 27$ at P. Find the coordinates of P.

4. The line from the point (2, 1) perpendicular to the line $2x + 3y - 6 = 0$ meets the x-axis at P and the y-axis at Q. Find the coordinates of P and Q, and the area enclosed by the axes and this line.

5. The points P, Q and R have coordinates (−2, 1), (4, 5) and (−1, 6) respectively. The line from R perpendicular to PQ meets PQ at N. Find the coordinates of N.

6. A parallelogram has vertices P (2, 3), Q (5, 7), R(x, y) and S (−2, 6). Find the coordinates of R.

7. Two points A and B have coordinates (1, 5) and (3, 3) respectively. The line through the midpoint of the line AB parallel to the line $4x - 3y - 12 = 0$ meets the x-axis at the point P and the y-axis at the point Q. Find the coordinates of P and Q, and the area enclosed by the axes and this line.

8.7 Reducing Non-Linear Graphs to Linear Form

Often we need to reduce a non-linear equation, such as $y = ax^3 + bx^2$ to the linear form $y = mx + c$ to enable us determine the values of the constants by calculating the gradient or the y-intercept.

Consider the equation $y = ax^n$.

This equation can be reduced to a linear form by taking logarithms of each side.

$$\log y = \log ax^n$$

$$\log y = \log a + n \log x$$

Writing log y as Y, log x as X and log a as c, the equation becomes $Y = c + mX$, which represents a straight line of gradient m. Thus, if we plot log y against log x we obtain a set of nearly collinear points. Then, we draw the line of best fit.

The gradient of the line determines the value of the constant n, and log a is found from the y-intercept.

Examples

1. The table shows experimental values of two variables x and y

x	1	2	3	4	5	6
y	5.75	22.9	51.3	91.2	144.5	208.9

It is know that x and y are related by the equation $y = ax^n$, where a and n are constants. Draw a suitable linear graph and use it to estimate the value of a and of n.

Taking logarithm of each side gives

$$\log y = \log a + n \log x$$

We construct a new table of values for this equation.

log x	0	0.30	0.47	0.60	0.70	0.78
log y	0.76	1.36	1.71	1.96	2.16	2.32

We plot the points shown in this table and connect them with a line, as shown in Figure 8.5.

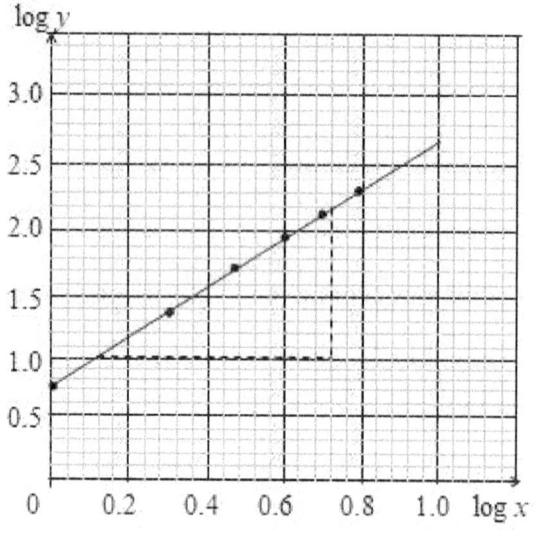

Figure 8.5

To find the gradient, we choose two points on the graph. Using the following points (0.12, 1) and (0.72, 2.2), we have

$$n = \frac{2.2 - 1}{0.72 - 0.12} = 2$$

The intercept on the vertical axis is 0.76.

Therefore $\qquad \log a = 0.76$

$$a = 5.75$$

2. The relation between x and y is given as $y = ax^2 + bx$, where a and b are constants. The table below gives the corresponding values of x and y.

x	1	2	3	4	5
y	5	16	33	56	85

Draw a suitable linear graph and use it to express y in terms of x. Estimate the value of y when x is 6.

$$y = ax^2 + bx$$

Dividing each side of the equation by x gives

$$\frac{y}{x} = ax + b$$

Reducing Non-Linear Graphs to Linear Form 119

We construct a new table of values for this equation.

x	1	2	3	4	5
y/x	5	8	11	14	17

Plot these points and connect them with a line as shown in Figure 8.6.

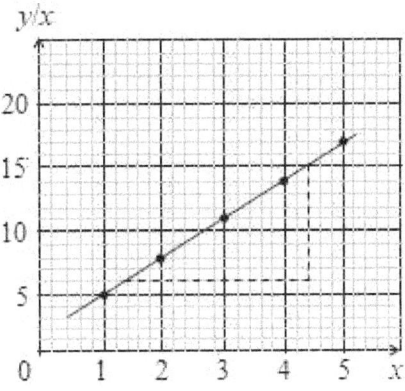

Figure 8.6

The gradient is

$$a = \frac{15 - 6}{4.4 - 1.4} = 3$$

An alternative way of finding the constant b is as follows. We substitute into the equation

$$\frac{y}{x} = 3x + b$$

the coordinates of a convenient point on the line. Substituting the coordinates of the point (3.6, 13) we have

$$13 = 3(3.6) + b$$

$$b = 2.2$$

Therefore the relation is $y = 3x + 2.2$.

When $x = 6$,

$$y = 3(6) + 2.2 = 20.2.$$

CHAPTER 8 Straight Line Graphs

Exercise 8.7

1. The data below fits an equation of the form $y = ax^n$.

x	2	3	4	5
y	22.4	50.4	89.6	140.0

   Draw a suitable linear graph and use it to find the value of a and of n.

2. The table below gives the corresponding values of two related variables x and y

x	1	2	3	4	5
y	6	12	24	48	96

   The relation between x and y is of the form $y = ab^x$, where a and b are constants. Draw a suitable linear graph and use it to express y in terms of x.

3. The relation between x and y is given as $y = ax^n$, where a and n are constants. The table below gives the corresponding values of x and y.

x	2	3	4	5	6
y	3.75	1.67	0.94	0.60	0.42

   By drawing a suitable linear graph estimate the value of a and of n. Find x when y is 2.

4. By using the given table of values draw a graph and find the relationship between the variables x and y in the form $y = ax^n$.

x	2	3	4	5	6
y	12.9	26.0	40.0	63.2	73.5

   Estimate the value of y when x is 7.

5. The table shows experimental values of two variables x and y.

x	0	1	2	3	4	5
y	8	9.5	14	21.5	32	45.5

   It is known that x and y are related by the equation $y = ax^2 + b$, where a and b are constants. By drawing a suitable linear graph estimate the value of a and of b. Find x when y is 35.

6. The table shows experimental values of two variables x and y

x	2	3	4	5	6
y	9.2	10.2	8.8	5.0	−1.2

Reducing Non-Linear Graphs to Linear Form

It is known that x and y are related by the equation $y = ax^2 + bx$, where a and b are constants. Using graph paper, plot y/x against x for the above data and use your graph to estimate the value of a and of b.

7. The table below gives the values of two related variables x and y.

x	0.2	0.4	0.6	0.8	1.2
y	15	9	7	6	5

The relation between x and y is of the form $y = a + b/x$, $x \neq 0$ and a and b are constants. Draw a suitable linear graph and use it to estimate the value of a and of b. Estimate the value of y when x is 0.5.

8. The table shows experimental values of two variables x and y.

x	1	2	3	4	5
y	1.65	1.50	1.75	2.10	2.49

It is known that x and y are related by the equation $y = ax + b/x$, $x \neq 0$ where a and b are constants. Draw a straight line graph and use it to estimate the value of a and of b, correct to two significant figures. Estimate the value of y when x is 8.

9. The table shows experimental values of two variables x and y.

x	1	2	3	4
y	4.30	0.80	−0.97	−1.70

It is known that x and y are related by the equation $y = a/x^2 + bx$, $x \neq 0$ where a and b are constants. Using graph paper, plot x^2y against x^3 for the above data and use your graph to estimate the value of a and of b.

Review Exercise 8

1. Find the distance between the following pairs of points.

   (a) (4, 8), (5, 6)
   (b) (6, 3), (9, 4)
   (c) (2, 4), (6, 7)
   (d) (1, −4), (3, −7)
   (e) (1, 3), (3, −2)
   (f) (2, 9), (14, 4)

2. Find the coordinates of the midpoint of lines joining the following pairs of points.

   (a) (−1, 1), (5, 9)
   (b) (−2, −1), (8, 3)

(c) (2, 7), (7, 4) (d) (3, 8), (10, 0)

(e) (8, 3), (5, 2) (f) (−1, 6), (−3, 1)

3. Find the gradients of the lines joining the following pairs of points.

   (a) (1, 1), (8, 6) (b) (−1, −2), (−4, −6)

   (c) (7, 1), (4, −5) (d) (3, −4), (6, −4)

   (e) (−1, 1), (2, 5) (f) (−2, 4), (3, 1)

4. Find the equation of the lines passing through the given pairs of points.

   (a) (1, 4), (5, 6) (b) (2, 6), (4, 1)

   (c) (2, 2), (−1, −2) (d) (−2, 4), (1, 8)

   (e) (3, −2), (7, −5) (f) (4, −1), (5, 2)

5. Find the equation of the line through each point parallel to the given line.

   (a) $x - 4y - 4 = 0$, (3, 2)

   (b) $2x + 5y - 3 = 0$, (−2, 1)

   (c) $3x - 2y + 10 = 0$, (2, −1)

   (d) $3x + 2y + 8 = 0$, (3, 4)

6. Find the equation of the line through each point perpendicular to the given line.

   (a) $3x + 2y - 2 = 0$, (−3, 2)

   (b) $x - 4y + 2 = 0$, (2, 3)

   (c) $2x - 6y + 15 = 0$, (−2, −3)

   (d) $8x + 6y - 3 = 0$, (1, −2)

7. Find the points of intersection of each of the following pairs of lines.

   (a) $5x + 2y = 7$, $3x - y = 13$

   (b) $2x + 3y - 8 = 0$, $3x + 4y - 13 = 0$

(c) $x - 3y - 2 = 0$, $3x - 7y - 4 = 0$

(d) $2x - y - 9 = 0$, $3x + 4y + 14 = 0$

8. Prove that the polygon with vertices (3, 2), (0, 5), (−3, 2) and (0, −1) is a rhombus.

9. Prove that (−2, 4), (10, 9), (15, −3) and (3, −8) are the vertices of a square.

10. For the triangle with vertices A (2, 3), B (4, 7), C (−6, −5) verify that the line joining the midpoints of two sides is parallel to the third side.

11. The vertices of a quadrilateral ABCD are A (1, 4), B (2, −7), C (−3, −2) and D (−4, 3).

    (a) Find the midpoint of the sides

    (b) Prove that the midpoints of the sides are vertices of a parallelogram.

12. Find an equation of the line parallel to $y = 3x + 2$ having the same x-intercept as the line $4x + 3y = 8$.

13. Find an equation of the line perpendicular to $y = -3x + 4$ having the same y-intercept as the line $5x - 3y = 6$.

14. Find an equation of the line containing the point (−1, 3) and the point of interception of the lines $x - 4y = 5$ and $x - 2y = 4$.

15. The table below gives the corresponding values of two variables x and y for the relation $y = ab^x$, where a and b are constants.

x	1	2	3	4	5	6
y	4.50	7.62	10.13	13.20	22.78	36.40

    By means of a suitable graph, estimate correct to one decimal place

    (a) the value of a and of b

    (b) the value of y when x = 4.5.

16. The table below gives the corresponding values of two related variables x and y.

x	1.5	2.0	2.5	3.0	3.5
y	3.6	5.6	10.0	17.4	19.6

   The relation between x and y is of the form $y = ax^n$, where a and n are constants. Draw a suitable linear graph and use it to estimate, correct to two decimal places, the value of a and of n.

17. The table shows experimental values of two variables x and y

x	2	4	6	8
y	12.4	48.8	133.2	289.6

    It is known that x and y are related by the equation $y = ax^3 + bx$, where a and b are constants.

    (a) Draw a suitable linear graph and use it to estimate, correct to one decimal place, the value of a and of b.

    (b) Estimate the value of y when $x = 2.4$.

9 Circular Measure

One way to measure angle is in radians.

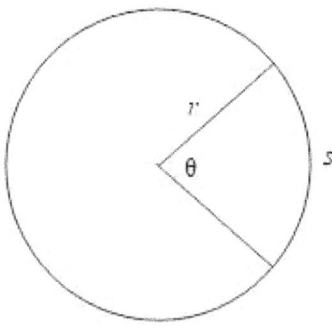

Figure 9.1

The angle θ whose vertex is the centre of the circle, cuts an arc of length s, as shown in Figure 9.1. We define the radian measure on a circle of any radius as follows:

Radian measure of θ is

$$\frac{\text{Length of arc}}{\text{Radius}} = \frac{s}{r}$$

When the length of the arc is equal in length to the radius of the circle then θ is 1 radian.

The length of the circumference of any circle is given by the formula $C = 2\pi r$

Therefore, the angle representing one complete revolution of a circle is

$$2\pi r/r = 2\pi \text{ radian.}$$

It follows that 2π radian = 360°, and π radian = 180°.

Dividing both sides of π radian = 180° by π gives

$$1 \text{ radian} = \frac{180}{\pi} \approx 57.3°$$

Also, $1° = \frac{\pi}{180} \approx 0.0175$ radian

Converting Degrees to Radians

To convert degrees to radians, multiply by π/180.

Examples

Convert the following to radians.

1. 135°

$$135° = 135 \times \frac{\pi}{180}$$

$$= \frac{3}{4}\pi \text{ radians}$$

2. 315°

$$315° = 315 \times \frac{\pi}{180}$$

$$= \frac{7}{4}\pi \text{ radians}$$

Often when we convert degrees to radians we give the answer in multiples of π.

Converting Radians to degrees

To convert radians to degree, multiply by 180/π

Examples

Convert the following to degrees.

1. ¼ π

$$\frac{1}{4}\pi = \frac{1}{4}\pi \times \frac{180}{\pi}$$

$$= 45°$$

2. $\frac{3}{5}\pi$

$$\frac{3}{5}\pi = \frac{3}{5}\pi \times \frac{180}{\pi}$$

$$= 108°$$

Exercise 9.1

1. Convert the following to radians.

   (a) 30°   (b) 60°   (c) 90°

   (d) 120°   (e) 150°   (f) −210°

   (g) 225°   (h) 270°   (i) 405°

   (j) −240°   (k) −330°   (l) 300°

2. Convert the following to degrees.

   (a) $\pi/5$ (b) $\frac{3}{4}\pi$ (c) $7\pi/3$

   (d) 3π (e) $-5\pi/4$ (f) $4\pi/3$

   (g) $-7\pi/5$ (h) $-8\pi/3$ (i) $4\pi/9$

   (j) $\pi/10$ (k) -4π (l) $5\pi/2$

9.2 Length of an Arc and Area of a Sector of a Circle

Length of an arc of a Circle

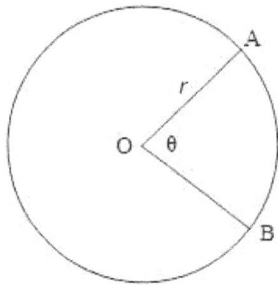

Figure 9.2

Figure 9.2 shows a circle of radius r with centre O. The arc AB subtends an angle θ radians at the centre. By the definition of radian measure

128 CHAPTER 9 Circular Measure

$$\frac{\text{Length of arc AB}}{r} = \theta$$

So length of arc AB = $r\theta$

Generally, the length of an arc which subtends θ radians at the centre of a circle radius r is $r\theta$.

Examples

1. Find the length of an arc which subtends an angle of 3 radians at the centre of a circle radius 4.5 cm.

$$\text{Length of arc} = r\theta$$

$$= 4.5 \times 3$$

$$= 13.5 \text{ cm}$$

2. A circle has radius of 10 cm. An arc AB subtends an angle of $3\pi/5$ radians at the centre of the circle. Find the perimeter of the sector containing the angle.

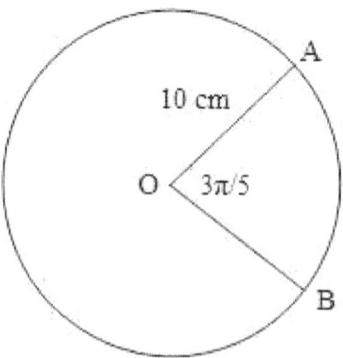

Length of arc

$$AB = 10 \times \frac{3}{5}\pi$$

$$= 18.85$$

The perimeter of the sector = 10 + 10 + 18.85

$$= 38.85 \text{ cm}$$

Area of a Sector of a Circle

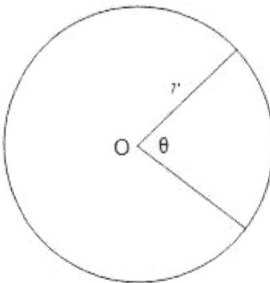

Figure 9.3

Figure 9.3 shows a circle of radius r. The area A of the sector containing angle θ is given by

$$A = \frac{1}{2}r^2\theta$$

where θ is measured in radians.

Examples

1. Find the area of the sector containing an angle of 150° in a circle of radius 12 cm.

    To use the formula $A = \frac{1}{2} r^2\theta$, first convert 150° to radians.

    $$150° = 150 \times \frac{\pi}{180}$$

    $$= \frac{5}{6}\pi$$

    $$\text{Area of sector} = \frac{1}{2} \times 12^2 \times \frac{5}{6}\pi$$

    $$= 188.5 \text{ cm}^2$$

2.

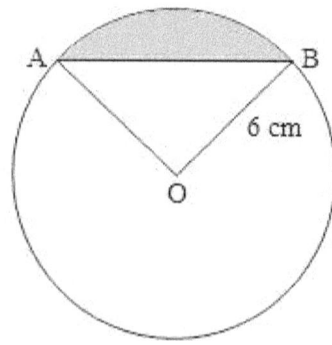

The diagram shows two points A and B on the circumference of a circle of radius 6 cm, centre O. Given that angle AOB is 2π/3 radians, find the area of the segment, shown shaded in the diagram.

$$\text{Area of sector AOB} = \tfrac{1}{2} r^2 \theta$$

$$= \tfrac{1}{2} \times 6^2 \times \tfrac{2}{3}\pi$$

$$= 37.7 \text{ cm}^2$$

$$\text{Area of triangle AOB} = \tfrac{1}{2} r^2 \sin \theta$$

$$= \tfrac{1}{2} \times 6^2 \times \sin \tfrac{2}{3}\pi$$

$$= 15.6 \text{ cm}^2$$

$$\text{Area of segment} = 37.7 - 15.6 = 22.1 \text{ cm}^2$$

Exercise 9.2

1. Find the length of the arc and the area of the circular sector whose radius and angle are:

   (a) 15 cm, 0.72 rad.
   (b) 8 m, 2.5 rad.
   (c) 6 m, 1.2 rad.
   (d) 12 cm, ⅔π rad.
   (e) 20 cm, 3π/2 rad.
   (f) 15 m, 150°

2. Find in radian, the angle subtended at the centre of a circle with the given radius r and arc length l.

(a) $r = 5$ cm, $l = 6$ cm (b) $r = 1.5$ cm, $l = 3.75$ cm

(c) $r = 21$ cm, $l = 44$ cm (d) $r = 2.5$ cm, $l = 2$ cm

3. The area of a sector of a circle radius 3 cm is 18 cm^2. Find the angle contained by the sector.

4. An arc of a circle subtends an angle of 1.5 radians at centre. Find the radius of the circle if the length of the arc is 9 cm.

5. An arc of length 7.2 cm subtends an angle of 1.5 radians at the centre of a circle, what is the radius of the circle?

6. An arc subtends an angle of 2 radians at the centre of a circle and a sector of area 49 cm^2 is bounded by this arc and two radii. Find the radius of the circle.

7. The arc of a sector in a circle radius 3 cm is 4.5 cm long. Find the area of the sector.

8. An arc AB of a circle radius 4.2 cm subtends an angle 150° at the centre O. Find the perimeter of the sector containing the angle.

9. A chord AB 18 cm long subtends an angle of 90° at the centre O of a circle. Find the perimeter of the sector AOB containing the angle, correct to one decimal place.

10. A chord AB subtends an angle 120° at the centre of a circle radius 15 cm. Find the perimeter of the minor segment cut off by AB, correct to one decimal place.

11. A chord subtends 1.5 radians at the centre of a circle radius 12 cm. Find the areas of the minor and major segments.

12. A chord AB divides a circle of radius 4 cm into two segments. If AB subtends an angle of 135° at the centre of the circle, find the area of the major segment.

13. A circle of radius 1.5 cm has a sector with area 3.6 cm^2. Calculate the perimeter of the sector.

14.

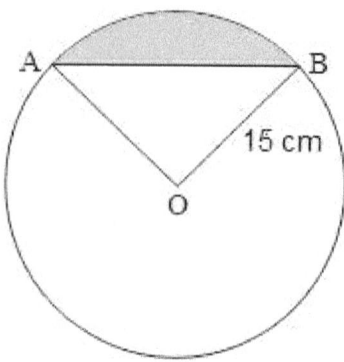

The diagram shows a circle with centre O and a chord AB. The radius of the circle is 15 cm and angle AOB is 1.2 radians.

(a) Find the area of the shaded region.

(b) Find the perimeter of the shaded region.

15.

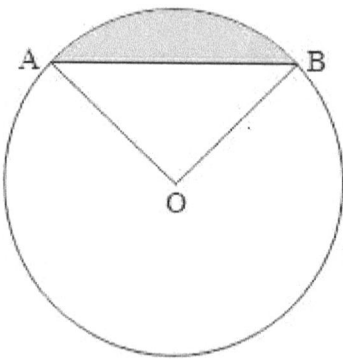

The diagram shows two points A and B on the circumference of a circle, centre O. Given that angle AOB is 1.5 radians, find the ratio of the area of the segment shown shaded in the diagram, to the area of the circle.

16. A chord AB subtends 120° at the centre O of a circle with radius is 10 cm. Find:

(a) the length of the minor arc AB

(b) the area of the triangle OAB

(c) the area of the minor segment cut off by AB.

Review Exercise 9

1. Convert the following angles to radians.

   (a) 495°   (b) −45°   (c) −150°

   (d) 600°   (e) −300°   (f) 324°

2. Convert the following angles to degrees.

   (a) $\pi/3$ rad.   (b) $3\pi/2$ rad.   (c) $-\pi/4$ rad.

   (d) 5π rad.   (e) $8\pi/5$ rad.   (f) $13\pi/15$ rad.

3. Find the length of the arc and the area of the circular sector of a circle with the given radius and angles.

   (a) 9 cm, 3 rad.   (b) 15 m, 2.5 rad.

   (c) 8.5 cm, $\frac{3}{4}\pi$ rad.   (d) 12 cm, $4\pi/5$ rad.

4. Find in radian the angle subtended at the centre of the circle with the given radius r and arc length l.

   (a) $r = 15$ cm, $l = 9$ cm

   (b) $r = 1.2$ m, $l = 3$ m

   (c) $r = 14$ cm, $l = 0.33$ m

   (d) $r = 0.24$ m, $l = 4.8$ cm

5. An arc of length 21.9 cm subtends an angle of 7.3 radians at the centre of a circle. Find the radius of the circle.

6. A wheel makes 12 revolutions a minute. Through how many radians does the wheel revolve in one minute?

7. The flywheel of an engine makes 35 revolutions in a second, how long will it take to turn through 10 radians? [Take $\pi = 22/7$]

8. The floodlights at a sport stadium spread its illumination over 0.75 radians to a distance of 50 metres. Find the maximum area that is floodlit.

9.

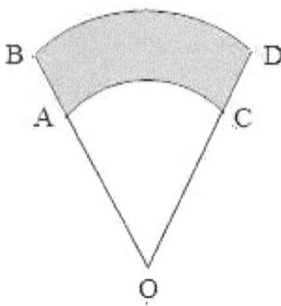

The diagram shows a sector of a circle centre O. OD = 5 cm, CD = 1.2 cm and angle AOC is 0.75 radians.

(a) Find the area of the shaded region.

(b) Find the percentage of the whole sector that the shaded area represents.

10.

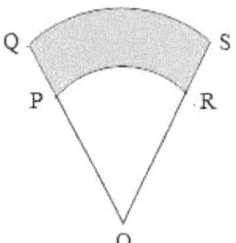

The diagram shows a sector of a circle centre O. OP = 4 cm and OQ = 6 cm. If the length of the arc PR is 7.2 cm, find:

(a) the size of angle POR in radians to 1 decimal place

(b) the perimeter of the shaded region

(c) Find the area of the shaded region

11.

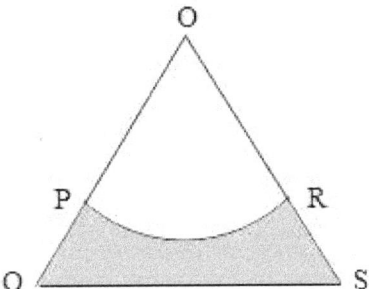

The diagram shows an isosceles triangle OQS in which OQ = OS = 16 cm and angle QOS = 1.2 radians. PR is an arc of a circle centre O and radius 8 cm.

(a) Find the area of the shaded region.

(b) Find the perimeter of the shaded region.

12. A chord AB subtends an angle of 2.1 radians at the centre O of a circle of radius 15 cm. Find:

(a) the length of the minor arc AB.

(b) the area of triangle OAB.

(c) the area of the major segment cut off by AB.

13.

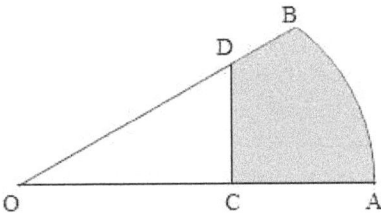

The diagram shows a sector AOB of a circle with centre O. Angle AOB = 0.8 radians. The point D lies on OB such that the length of DB is 4 cm.

The point C lies on OA such that DC is perpendicular to OA and DC = 1.5 cm.

(a) Find the length of OD and of OC.

(b) Find the perimeter of the shaded region.

(c) Find the area of the shaded region.

14. A chord AB divides a circle of radius 6 cm into two segments. If AB subtends an angle of $60°$ at the centre of the circle, find the area of the minor segment.

15. A chord divides a circle of radius 12 cm into two segments. If the chord subtends an angle of 1.5 radians at the centre of the circle, find the perimeter of the region bordered by the minor arc and the chord.

10 Trigonometry

10.1 The Trigonometric Functions

Three basic trigonometric functions are sine, cosine and tangent, written in brief as sin, cos and tan respectively.

Consider the circle shown in Figure 10.1 which has radius 1, called the unit circle.

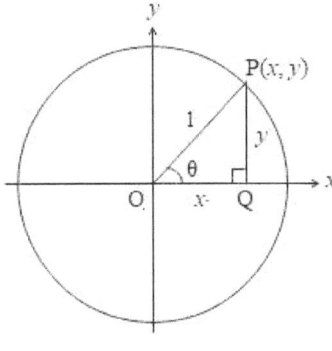

Figure 10.1

The point P having coordinates (x, y) is located on the terminal side of angle θ with its vertex at the origin. The line PQ is drawn perpendicular to the x-axis to form the right-angled triangle POQ. The length of OP is 1. The three trigonometric function of angle θ are defined as follows.

$$\sin \theta = \frac{y}{1} = y,$$

$$\cos \theta = \frac{x}{1} = x$$

and $\tan \theta = \frac{y}{x} = \frac{\sin \theta}{\cos \theta}$

Notice that $\cos \theta$ and $\sin \theta$ are the x- and y- coordinates respectively of a point on the unit circle. The x- and y- coordinates of any point on the unit circles can have values from -1 to 1. Hence, for any angle θ, we have $-1 \leq \sin \theta \leq 1$ and $-1 \leq \cos \theta \leq 1$.

There are three other trigonometric functions: cotangent (cot), secant (sec) and cosecant (cosec) defined as follows:

$$\cot \theta = \frac{x}{y} = \frac{1}{\tan \theta} \qquad \sec \theta = \frac{1}{x} = \frac{1}{\cos \theta} \quad \text{and} \quad \csc \theta = \frac{1}{y} = \frac{1}{\sin \theta}$$

138 CHAPTER 10 Trigonometry

These identities are undefined when the denominator is zero.

Evaluating Trigonometric Functions of 0°, 90°, 180°, 270° and 360°

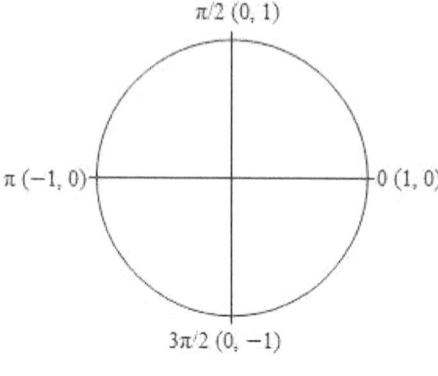

Figure 10.2

Figure 10.2 shows the coordinates of four points on a unit circle. We can use the coordinates to evaluate the trigonometric functions of their corresponding angles.

The point (1, 0) gives $x = 1$ and $y = 0$. Then

$$\sin 0° = \frac{0}{1} = 0, \qquad \cos 0° = \frac{1}{1} = 1 \qquad \tan \theta° = \frac{0}{1} = 0$$

$$\operatorname{cosec} \theta° = \frac{1}{0} \qquad \sec 0° = \frac{1}{1} = 1 \qquad \cot \theta° = \frac{1}{0}$$

The values of cot 0° and cosec 0° are undefined because the denominator is 0.

The results summarized in the following table are obtained in a similar way.

θ	$\sin \theta$	$\cos \theta$	$\tan \theta$	$\cot \theta$	$\sec \theta$	$\operatorname{cosec} \theta$
0	0	1	0	undefined	1	undefined
$\pi/2$	1	0	undefined	0	undefined	1
π	0	−1	0	undefined	−1	undefined
$3\pi/2$	−1	0	undefined	0	undefined	−1
2π	0	1	0	undefined	1	undefined

Special Angles

The values of trigonometric functions of angles, called special angles, can be found exactly using the aid of two kinds of right-angled triangles.

Trigonometric Functions of 45°

Consider the right-angled triangle shown in Figure 10.3 in which the two sides have length 1.

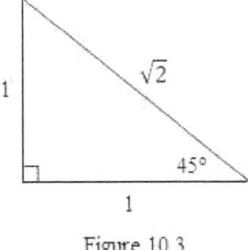

Figure 10.3

Using the Pythagoras' theorem we find the length of the hypotenuse as

$$\sqrt{1^2 + 1^2} = \sqrt{2}.$$

From the diagram, we see that:

$$\sin 45° = \frac{1}{\sqrt{2}} = \frac{\sqrt{2}}{2}$$

$$\cos 45° = \frac{1}{\sqrt{2}} = \frac{\sqrt{2}}{2}$$

and $\tan 45° = 1$

Trigonometric Functions of 30° and 60°

Consider the equilateral triangle shown in Figure 10.4 in which all sides have length 2.

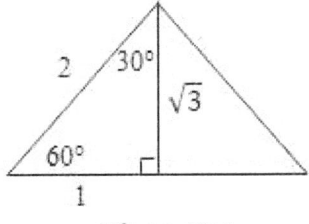

Figure 10.4

Recall from plain geometry that any line drawn from a vertex of an equilateral triangle perpendicular to a base bisects the angle at the vertex and it bisects the base. Using the Pythagoras' theorem we find the length of the altitude as $\sqrt{2^2 - 1^2} = \sqrt{3}$. From Figure 10.4 you can see that:

$$\sin 60° = \frac{\sqrt{3}}{2} \qquad \sin 30° = \frac{1}{2}$$

$$\cos 60° = \frac{1}{2} \qquad \cos 30° = \frac{\sqrt{3}}{2}$$

and $\quad \tan 60° = \sqrt{3} \qquad \tan 30° = \frac{1}{\sqrt{3}} = \frac{\sqrt{3}}{3}$

Trigonometric Functions of any Angle

The definitions of trigonometric functions can be extended to cover any angle.

Consider the point (x, y) on the terminal side of angle θ, as shown in Figure 10.5. By definition of the trigonometric functions, we know that:

$$\sin \theta = \frac{y}{r} \qquad \cos \theta = \frac{x}{r} \quad \text{and} \quad \tan \theta = \frac{y}{x}$$

In the first quadrant x and y are positive. Since r is always positive, the signs of the functions depends on the signs of x and y. Therefore, $\sin \theta$, $\cos \theta$ and $\tan \theta$ are positive.

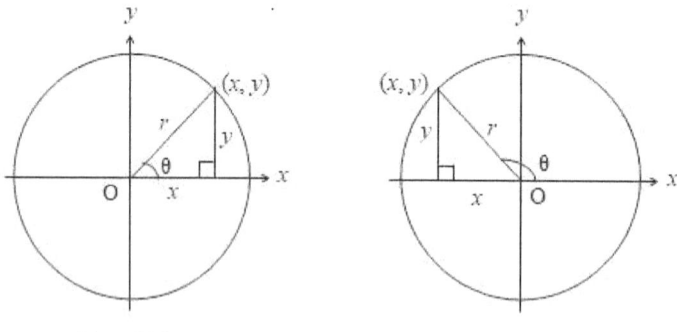

Figure 10.5   Figure 10.6

Figure 10.6 shows θ in the second quadrant. In this quadrant, y is positive and x is negative. Therefore, $\sin \theta$ is positive, $\cos \theta$ and $\tan \theta$ are negative.

By similar reasoning, you can determine the signs of the trigonometric functions in the rest of the quadrants. The following table shows the signs of the trigonometric functions in the quadrants.

Quadrant	x	y	sin θ = y/r	cos θ = x/r	tan θ = y/x
First	+	+	+	+	+
Second	−	+	+	−	−
Third	−	−	−	−	+
Fourth	+	−	−	+	−

Observe that all the trigonometric functions are positive in the first quadrant. Sine, tangent and cosine are positive in the second, third and fourth quadrant respectively.

The signs of cosec θ, sec θ and cot θ are the same as that of sin θ, cos θ and tan θ respectively.

You can see from Figure 10.7 that the terminal side of an angle greater than 90° forms a corresponding acute angle with the x-axis.

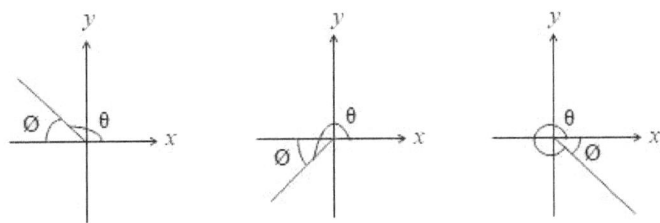

Figure 10.7

In each quadrant, the trigonometric functions of θ and Ø are equal except possibly in sign.

Examples

Evaluate each of the following trigonometric functions

1. $\sin\dfrac{5}{6}\pi$

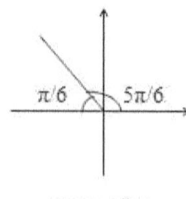

Figure 10.8

The angle 5π/6 lies in the second quadrant. The corresponding acute angle is π − 5π/6 = π/6, as shown in Figure 10.8. The sine function is positive in the second quadrant, so

$$\sin\frac{5\pi}{6} = \sin\frac{\pi}{6}$$
$$= \frac{1}{2}$$

2. cos 225°

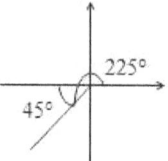

Figure 10.9

The angle 225° lies in the third quadrant. The corresponding acute angle is 225° − 180° = 45°, as shown in Figure 10.9. The cosine function is negative in the third quadrant, so

$$\cos 225° = -\cos 45°$$
$$= -\frac{\sqrt{2}}{2}$$

3. $\tan\dfrac{5\pi}{3}$

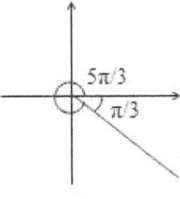

Figure 10.10

The angle 5π/3 lies in the fourth quadrant. The corresponding acute angle is 2π − 5π/3 = π/3, as shown in Figure 10.10. The tangent function is negative in the fourth quadrant, so

$$\tan\frac{5\pi}{3} = -\tan\frac{\pi}{3}$$
$$= -\sqrt{3}$$

4. cos 780°

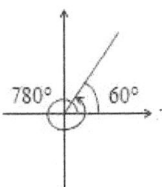

Figure 10.11

The angle 780° is co-terminal with 60°, as shown in Figure 10.11. The cosine function is positive in the first quadrant, so

$$\cos 780° = \cos 60°$$
$$= \frac{1}{2}$$

Trigonometric Functions of Negative Angles

When we rotate anticlockwise, the angle is positive while a clockwise rotation gives a negative angle.

144 CHAPTER 10 Trigonometry

An angle θ is measured clockwise from the x-axis, as shown in Figure 10.12. The point (x, y) lies on the terminal side of θ.

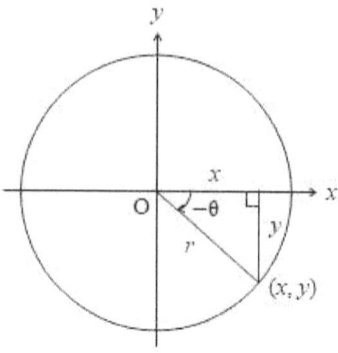

Figure 10.12

Using the definition of trigonometric functions we obtain the following results.

$\sin(-\theta) = \dfrac{-y}{r} = -\sin\theta$

$\operatorname{cosec}(-\theta) = \dfrac{r}{-y} = -\operatorname{cosec}\theta$

$\cos(-\theta) = \dfrac{x}{r} = \cos\theta$

$\sec(-\theta) = \dfrac{r}{x} = \sec\theta$

$\tan(-\theta) = \dfrac{-y}{x} = -\tan\theta$

$\cot(-\theta) = \dfrac{x}{-y} = -\cot\theta$

By similar reasoning you can obtain the signs of trigonometric functions of negative angles in the remaining quadrants.

Examples

Evaluate the following trigonometric functions.

1. tan(− 150°)

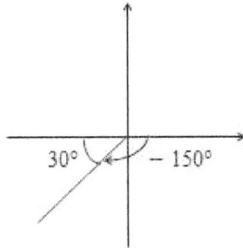

Figure 10.13

The angle $-150°$ lies in the third quadrant, as shown in Figure 10.13. $-150°$ is co-terminal with the angle $210°$. The corresponding acute angle of both $-150°$ and $210°$ is $30°$. Because tangent is positive in the third quadrant, we have

$$\tan(-150°) = \tan 30°$$
$$= \frac{\sqrt{3}}{3}$$

2. $\cos(-4\pi/3)$

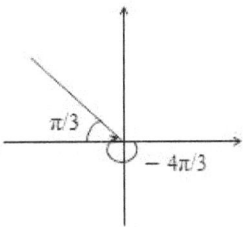

Figure 10.14

The angle $-4\pi/3$ lies in the second quadrant, as shown in Figure 10.14. $-4\pi/3$ is co-terminal with the angle $2\pi/3$. The corresponding acute angle of both $-4\pi/3$ and $2\pi/3$ is $\pi/3$. Because cosine is negative in the second quadrant we have

$$\cos\left(-\frac{4\pi}{3}\right) = -\cos\frac{\pi}{3}$$
$$= -\frac{1}{2}$$

CHAPTER 10 Trigonometry

Exercise 10.1

1. Find exact values of the following functions.

   (a) cos 150°   (b) tan 120°   (c) sin 210°

   (d) cot 225°   (e) sin 135°   (f) cosec 315°

   (g) sec(− 30°)   (h) cosec(− 120°)   (i) cot(− 135°)

   (j) tan(− 60°)   (k) cos 405°   (l) sin 390°

   (m) sec 600°   (n) tan 765°   (o) sin(− 540°)

2. Find exact values of the following functions.

   (a) cos(π/2)   (b) sin(π/4)   (c) tan(π/6)

   (d) cosec(π/3)   (e) sec(7π/6)   (f) cot(3π/4)

   (g) tan(2π/3)   (h) sin(3π/2)   (i) cos(2π/3)

   (j) sec(5π/3)   (k) cot(− 4π/3)   (l) cosec(− 5π/4)

   (m) cos(− 16π/3)   (n) sin(− 9π/4)   (o) tan(− 5π/3)

10.2 Trigonometric Identities

An identity is an equation that is true for every real number in the domain of the variable.

Consider the unit circle given by $x^2 + y^2 = 1$. A point having coordinates (cos θ, sin θ) on the circle corresponds to an angle θ. It follows that

$$\cos^2 \theta + \sin^2 \theta = 1 \qquad (1)$$

Dividing (1) by $\cos^2 \theta$ gives

$$1 + \tan^2 \theta = \sec^2 \theta \qquad (2)$$

Again dividing (1) by $\sin^2 \theta$ gives

$$1 + \cot^2 \theta = \text{cosec}^2 \theta \qquad (3)$$

Trigonometric Identities

The following identities are listed for easy reference.

1. $\cot\theta = 1/\tan\theta \quad \sec\theta = 1/\cos\theta \quad \csc\theta = 1/\sin\theta$

2. $\tan\theta = \sin\theta/\cos\theta \quad \cot\theta = \cos\theta/\sin\theta$

3. $\sin^2\theta + \cos^2\theta = 1 \quad 1 + \tan^2\theta = \sec^2\theta \quad 1 + \cot^2\theta = \csc^2\theta$

Examples

Simplify:

1. $\sec^2\theta - \tan^2\theta$

Begin by converting the ratios to equivalent ratios in sine and cosine.

$$\sec^2\theta - \tan^2\theta = \frac{1}{\cos^2\theta} - \frac{\sin^2\theta}{\cos^2\theta}$$

$$= \frac{1-\sin^2\theta}{\cos^2\theta}$$

$$= \frac{\cos^2\theta}{\cos^2\theta}$$

$$= 1$$

Note that $\sin^2\theta + \cos^2\theta = 1$ gives $\sin^2\theta = 1 - \cos^2\theta$ and $\cos^2\theta = 1 - \sin^2\theta$.

2. $\dfrac{\cos\theta}{1-\sin\theta} + \dfrac{\cos\theta}{1+\sin\theta}$

$$\frac{\cos\theta}{1-\sin\theta} + \frac{\cos\theta}{1+\sin\theta} = \frac{\cos\theta(1+\sin\theta) + \cos\theta(1-\sin\theta)}{(1-\sin\theta)(1+\sin\theta)}$$

$$= \frac{2\cos\theta}{1-\sin^2\theta}$$

$$= \frac{2}{\cos\theta}$$

$$= 2\sec\theta$$

Exercise 10.2(a)

Simplify:

1. $\sin\theta \cot\theta$ 2. $\sec\theta \cot\theta$ 3. $(\sec\theta - 1)(\sec\theta + 1)$

4. $\dfrac{\sec^2\theta - 1}{1-\cos^2\theta}$ 5. $\dfrac{\sin^2\theta}{1-\cos\theta}$ 6. $\dfrac{\cot^2\theta}{\csc\theta - 1}$

7. $\dfrac{1-\cos^2\theta}{1-\sin^2\theta}$ 8. $\dfrac{\cos\theta}{\sin\theta \cot^2\theta}$ 9. $(1-\sin^2\theta)\sec^2\theta$

10. $\dfrac{\cos^2\theta}{1+\sin\theta} + \dfrac{\cos^2\theta}{1-\sin\theta}$ 11. $\cot\theta + \dfrac{\sin\theta}{1+\cos\theta}$ 12. $(\sin\theta + \cos\theta)^2 - 1$

Proving Trigonometric Identities

The general method of procedure used in proving trigonometric identities is to choose one side of the equation and use the fundamental trigonometric identities to simplify it until you obtain the expression on the other side. Occasionally, it would be convenient to reduce both sides to identical form.

Examples

Proof the following identities.

1. $\tan\theta + \cot\theta = \dfrac{1}{\sin\theta \cos\theta}$

We start with the expression on left side and convert it to the form given on the right side.

$$\tan\theta + \cot\theta = \dfrac{\sin\theta}{\cos\theta} + \dfrac{\cos\theta}{\sin\theta}$$

$$= \dfrac{\sin^2\theta + \cos^2\theta}{\sin\theta \cos\theta}$$

$$= \dfrac{1}{\sin\theta \cos\theta}$$

The identity is proved.

Another way to prove this identity is to take the right side.

$$\dfrac{1}{\sin\theta \cos\theta} = \dfrac{\sin^2\theta + \cos^2\theta}{\sin\theta \cos\theta}$$

$$= \dfrac{\sin^2\theta}{\sin\theta \cos\theta} + \dfrac{\cos^2\theta}{\sin\theta \cos\theta}$$

$$= \dfrac{\sin\theta}{\cos\theta} + \dfrac{\cos\theta}{\sin\theta}$$

$$= \tan\theta + \cot\theta$$

Trigonometric Identities 149

2. $\dfrac{\tan\theta}{\sec\theta-1} - \dfrac{\tan\theta}{\sec\theta+1} = 2\cot\theta$

$$\dfrac{\tan\theta}{\sec\theta-1} - \dfrac{\tan\theta}{\sec\theta+1} = \dfrac{\tan\theta(\sec\theta+1)-\tan\theta(\sec\theta-1)}{(\sec\theta-1)(\sec\theta+1)}$$

$$= \dfrac{2\tan\theta}{\sec^2\theta-1}$$

$$= \dfrac{2\tan\theta}{\tan^2\theta}$$

$$= \dfrac{2}{\tan\theta}$$

$$= 2\cot\theta$$

3. $\dfrac{\tan^2\theta}{1+\sec\theta} = \dfrac{1-\cos\theta}{\cos\theta}$

Here we take each side separately to obtain one common form equivalent to both sides.

We work with the left side first.

$$\dfrac{\tan^2\theta}{1+\sec\theta} = \dfrac{\sec^2\theta-1}{1+\sec\theta}$$

$$= \dfrac{(\sec\theta-1)(\sec\theta+1)}{1+\sec\theta}$$

$$= \sec\theta - 1$$

Next, we work with the right side.

$$\dfrac{1-\cos\theta}{\cos\theta} = \dfrac{1}{\cos\theta} - 1$$

$$= \sec\theta - 1$$

The identity is proved.

Exercise 10.2(b)

Prove the following identities.

1. $\cos\theta \operatorname{cosec}\theta = \cot\theta$

2. $(\sec^2\theta - 1)\cos^2\theta = \sin^2\theta$

3. $(1 - \cos^2\theta)\operatorname{cosec}^2\theta = 1$

4. $(1 - \cos^2\theta)\sec^2\theta = \tan^2\theta$

5. $(1+\tan^2\theta)\cos^2\theta = 1$ 6. $\sin^2\theta(1+\cot^2\theta) = 1$

7. $\dfrac{1}{\sec^2\theta} + \dfrac{1}{\csc^2\theta} = 1$ 8. $\sec\theta - \tan\theta\sin\theta = \cos\theta$

9. $(\cos\theta + \sin\theta)^2 - (\cos\theta - \sin\theta)^2 = 2$

10. $\dfrac{1}{1+\sin^2\theta} + \dfrac{1}{1+\csc^2\theta} = 1$

11. $(1+\tan\theta)^2 + (1-\tan\theta)^2 = 2\sec^2\theta$

12. $\dfrac{1}{1-\sin\theta} + \dfrac{1}{1+\sin\theta} = 2\sec^2\theta$

13. $(\tan\theta + \sec\theta)^2 = \dfrac{1+\sin\theta}{1-\sin\theta}$

14. $\dfrac{1-\tan^2\theta}{1+\tan^2\theta} = 1 - 2\sin^2\theta$

15. $\dfrac{2\tan\theta}{1+\tan^2\theta} = 2\sin\theta\cos\theta$

16. $\dfrac{\tan^2\theta}{1+\tan^2\theta} \times \dfrac{1+\cot^2\theta}{\cot^2\theta} = \sin^2\theta\sec^2\theta$

10.3 Solving Trigonometric Equations

A general method for solving trigonometric equations is shown in the following example.

Examples

1. Solve $4\sin x = 3\csc x$ for $0° < x < 360°$.

$$4\sin x = 3\csc x$$

$$4\sin x = \dfrac{3}{\sin x}$$

$$\sin^2 x = \dfrac{3}{4}$$

$$\sin x = \pm\dfrac{\sqrt{3}}{2}$$

$$\sin x = -\dfrac{\sqrt{3}}{2} \quad \text{and} \quad \sin x = \dfrac{\sqrt{3}}{2}$$

$$x = 240°, 300° \qquad\qquad x = 60°, 120°$$

2. Solve $\cot x + 2\cos x = 0$ for $0° < x < 360°$.

$$\cot x + 2\cos x = 0$$

$$\frac{\cos x}{\sin x} + 2\cos x = 0$$

$$\cos x + 2\sin x \cos x = 0$$

$$\cos x (1 + 2\sin x) = 0$$

$$\cos x = 0 \quad \text{and} \quad 1 + 2\sin x = 0$$

$$x = 90°, 270° \qquad\qquad \sin x = -\frac{1}{2}$$

$$x = 210°, 330°$$

3. Solve $2\sin x + \operatorname{cosec} x = 3$ for $0 < x < 2\pi$.

$$2\sin x + \operatorname{cosec} x = 3$$

$$2\sin x + \frac{1}{\sin x} = 3$$

$$2\sin^2 x - 3\sin x + 1 = 0$$

$$(2\sin x - 1)(\sin x - 1) = 0$$

$$2\sin x - 1 = 0 \quad \text{and} \quad \sin x - 1 = 0$$

$$\sin x = \frac{1}{2} \qquad\qquad \sin x = 1$$

$$x = \frac{\pi}{6}, \frac{5\pi}{6} \qquad\qquad x = \frac{\pi}{2}$$

Exercise 10.3

1. Solve the following equations for $0° < x < 360°$

   (a) $2\sin x = \operatorname{cosec} x$
   (b) $3\sec x = 4\cos x$
   (c) $\sqrt{3}\sin x - \cos x = 0$
   (d) $1 - 3\cot^2 x = 0$
   (e) $\operatorname{cosec}^2 x = 4\cot^2 x$
   (f) $\tan x = 2 - \cot x$

(g) $3\cos^2 x = \sin^2 x$

(h) $\sec^2 x + \tan^2 x = 7$

(i) $2\sin^2 x = 3\cos x$

(j) $2\sin^2 x + \cos x = 1$

2. Solve the following equations for $0 < x < 2\pi$.

(a) $\tan x = 2\sin x$

(b) $4\sin x = 3\operatorname{cosec} x$

(c) $\tan^2 x = \tan x$

(d) $\sin x = 4\operatorname{cosec} x + 3$

(e) $2\cos^2 x - \cos x - 1 = 0$

(f) $2\sin^2 x + 3\sin x + 1 = 0$

10.4 Graphs of Trigonometric Functions

The Basic Graph of Sine Function

The graph of the sine function shown in Figure 10.15 is called a sine curve.

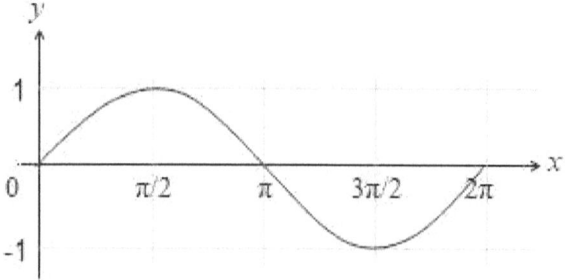

Figure 10.15

The graph represents one period of the function and is called one cycle of the sine curve. The sine function repeats itself after 2π in the positive and negative directions.

Graphs of Trigonometric Functions   153

The Basic Graph of the Cosine Function

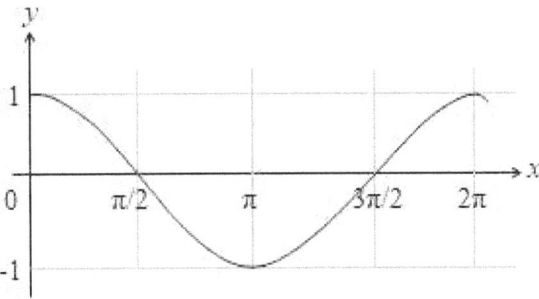

Figure 10.16

Figure 10.16 shows one cycle of the cosine curve. The cosine function repeats itself after 2π. Notice that you can obtain the graph of $y = \cos x$ if you shift the sine curve $\pi/2$ to left or right.

The Basic Graph of the Tangent Function

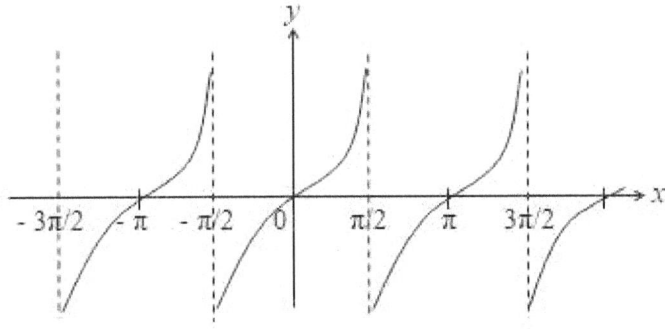

Figure 10.17

Figure 10.17 shows the graph of $y = \tan x$. Since $\tan x$ is undefined for $x = -3\pi/2, -\pi/2, \pi/2, 3\pi/2$ and so on, the graph has vertical asymptotes at these values. The tangent function repeats itself after π.

Amplitude

Recall that the maximum and minimum values of the sine and cosine functions are 1 and −1 respectively. The amplitude of the function $y = \sin x$ (or $y = \cos x$) is given by half the absolute value of the difference between the maximum and minimum values of the function. For instance, the amplitude of the function $y = \sin x$ is $|1 + 1|/2 = 1$.

The maximum and minmum values of $y = a \sin x$ is a and $-a$ respectively. The amplitude of the function is $|a + a|/2 = |a|$. Similarly, the amplitude of the function $y = a \cos x$ is $|a|$. For example, the amplitude of the function $y = -3\cos x$ is $|-3| = 3$.

Figure 10.18 shows the graphs of $y = \sin x$ and $y = 2 \sin x$.

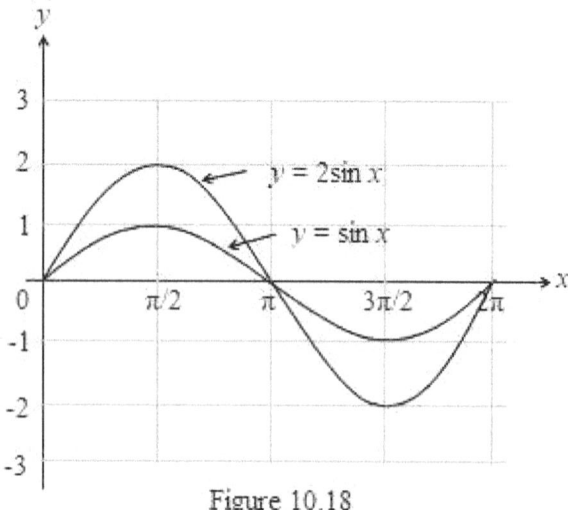

Figure 10.18

Observe that the graph of $y = 2 \sin x$ has the same shape as the graph of $y = \sin x$ and the y- values of $y = 2 \sin x$ have twice the values of those of $y = \sin x$.

Generally, the graph of $y = a \sin x$ (or $y = a \cos x$) is obtained by stretching the graph of $y = \sin x$ (or $y = \cos x$) along the y-axis by a factor of a.

Period

The values of the sine and cosine functions of any two angles that differ by 2π are equal. The sine and cosine functions are periodic with period 2π. The tangent function is periodic with a period of π.

Generally, the period of $y = \sin ax$ and $y = \cos ax$ is given by

$$\text{Period} = 2\pi/|a|$$

For example, the period of $y = \cos 6x$ is $2\pi/6 = \pi/3$ and the period of $y = \sin \frac{1}{2} x$ is $2\pi/(1/2) = 4\pi$.

Figure 10.19 shows the graphs of $y = \sin x$ and $y = \sin 2x$.

You can see from Figure 10.19 that $y = \sin x$ complete one cycle from $x = 0$ to $x = 2\pi$ and $y = \sin 2x$ complete one cycle from 0 to π. The graph of $y = \sin 2x$ is the same as $y = \sin x$ except that the period is different.

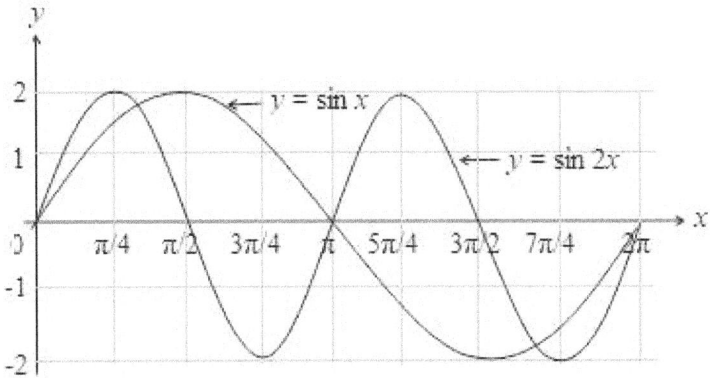

Figure 10.19

Generally, the graph of $y = \sin ax$ (or $y = \cos ax$) completes one cycle from $x = 0$ to $x = 2\pi/|a|$. If $0 < a < 1$, the period of $y = \sin ax$ is greater than 2π and represents a horizontal stretching of the graph of $y = \sin x$. Similarly, if $a > 1$, the period of $y = \sin ax$ is less than 2π and represents a horizontal shrinking of the graph of $y = \sin x$.

The graphs of $y = a \sin bx + c$ and $y = a \cos bx + c$ can be obtained from the graphs of $y = a \sin bx$ and $y = a \cos bx$ respectively by a vertical shift of c units. If $c > 0$ the graph is shifted upward and if $c < 0$, it is shifted downward.

Sketching the Graphs Sine and Cosine Functions

To sketch the graphs of the sine or cosine functions, it helps to find the following five key points in one period of the graph: the intercepts, maximum points, and minimum points. In additional to finding the key points, you will need to identify the amplitude of the function and find the period.

Examples

1. Sketch the graph of $y = \sin 2x + 3$ over one period.

   The amplitude is 1. Because $b = 2$, the period is $2\pi/2 = \pi$. Divide the interval $0 \leq x \leq \pi$ into four equal parts with x-values: 0, $\pi/4$, $\pi/2$, $3\pi/4$ and π. The five key points on the graph of $y = \sin 2x$ are:

   (0, 0)   ($\pi/4$, 1)   ($\pi/2$, 0)   ($3\pi/4$, -1) and (π, 0)

   The graph of $y = \sin 2x + 3$ is the graph of $y = \sin 2x$ shifted 3 units upward. It follows that the key points on this graph are:

   (0, 3)   ($\pi/4$, 4)   ($\pi/2$, 3)   ($3\pi/4$, 2) and (π, 3)

   Note that, you can obtain the same results when you evaluate $y = \sin 2x + 3$ at $x = 0$, $\pi/4$, $\pi/2$, $3\pi/4$ and π.

   Plot the points and connect them with a smooth curve to obtain the graph shown in Figure 10.20.

   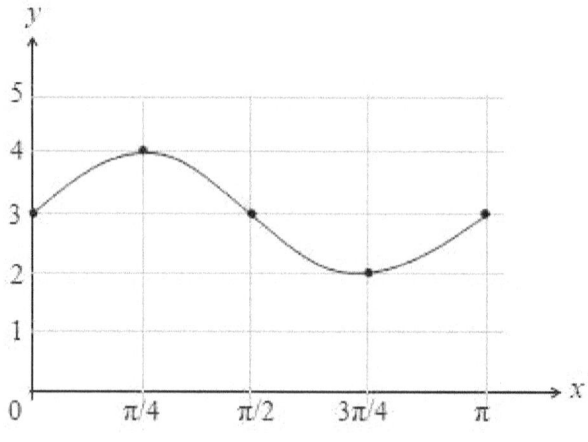

   Figure 10.20

2. Sketch the graph of $y = 2 \cos 3x$

   The ampplitude is 2 and the period is $2\pi/3$. Divide the interval $0 \leq x \leq 2\pi/3$ into four equal parts with x-values: 0, $\pi/6$, $\pi/3$, $\pi/2$ and $2\pi/3$. The five key points are: (0, 2), ($\pi/6$, 0), ($\pi/3$, -2), ($\pi/2$, 0) and ($2\pi/3$, 2).

The graph is shown below.

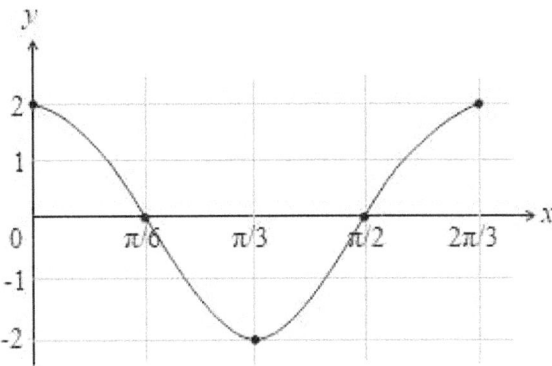

Exercise 10.4

1. Write down the amplitude and period of the following functions.

   (a) $3 \sin 4x$ (b) $-2 \cos 3x$

   (c) $2 \sin 5x$ (d) $-5 \sin \tfrac{1}{2} x$

   (e) $-\cos 6x$ (f) $4 \cos \tfrac{3}{4} x$

2. Sketch for $0 \le x \le 2\pi$, the graphs of the following functions.

   (a) $y = 2 \cos x$ (b) $y = \tfrac{1}{2} \sin x$

   (c) $y = 3 \cos 2x$ (d) $y = 2 \sin \tfrac{1}{3} x$

   (e) $y = 3 \cos \tfrac{1}{2} x$ (f) $y = 4 \sin \tfrac{1}{2} x$

3. Sketch for $0 \le x \le \pi$, the graphs of $y = \cos 2x$ and $y = 1 + \sin 2x$ on the same axes. Use your graph to solve the equation $\cos 2x - \sin 2x = 1$.

4. Sketch for $0 \le x \le \pi$, the graph of $y = 3 \sin 2x - 1$. Use the graph to find the coordinates of the maximum point.

5. Sketch for $0 \le x \le \pi$, the graphs of $y = \cos 2x$ and $y = \sin 2x$ on the same axes. Use your graph to solve the equation $\tan 2x - 1 = 0$.

Review Exercise 10

1. Find the exact value of each of the following functions.

   (a) $\tan 135°$ (b) $\sin 240°$ (c) $\cos(-150°)$

   (d) $\cot 120°$ (e) $\operatorname{cosec}(-210°)$ (f) $\sec 225°$

2. Simplfy:

   (a) $\sin\theta(\operatorname{cosec}\theta - \sin\theta)$ (b) $(\operatorname{cosec}\theta - 1)(\operatorname{cosec}\theta + 1)$

   (c) $\dfrac{\sin^2\theta}{1+\cos\theta} + \dfrac{\sin^2\theta}{1-\cos\theta}$ (d) $\dfrac{1}{\cos^2\theta} - \dfrac{1}{\cot^2\theta}$

   (e) $\dfrac{1}{\tan\theta} + \dfrac{\sin\theta}{1+\cos\theta}$ (f) $\dfrac{1+\sin\theta}{\cos\theta} + \dfrac{\cos\theta}{1+\sin\theta}$

3. Prove the following identities.

   (a) $(1 - \sin^2\theta)\sec^2\theta = 1$ (b) $(1 - \sin^2\theta)\operatorname{cosec}^2\theta = \cot^2\theta$

   (c) $\dfrac{1}{1-\cos^2\theta} - \dfrac{1}{\sec^2\theta - 1} = 1$ (d) $\dfrac{1+\cot^2\theta}{\operatorname{cosec}^2\theta - 1} = \sec^2\theta$

   (e) $\cot\theta + \dfrac{\sin\theta}{1+\cos\theta} = \operatorname{cosec}\theta$ (f) $\dfrac{\tan\theta}{\sec\theta - 1} + \dfrac{\tan\theta}{\sec\theta + 1} = 2\operatorname{cosec}\theta$

4. Solve the following equations for $0 < x < 2\pi$.

   (a) $2\sin x - \operatorname{cosec} x = 0$ (b) $\sec x = 2\cos x$

   (c) $3\cot x = \tan x$ (d) $\tan^2 x = 3 - 2\sec^2 x$

5. Write down the amplitude and period of the following functions.

   (a) $5\sin 2x$ (b) $-3\cos 6x$ (c) $2\sin 3x$

   (d) $-2\cos 4x$ (e) $4\cos\tfrac{1}{2}x$ (f) $-6\sin\tfrac{2}{3}x$

6. Sketch the graph of the following functions over one period.

   (a) $y = 2 + 3\sin 2x$ (b) $y = 2\cos 3x - 1$

7. Sketch for $0 \leq x \leq \pi$, the graphs of $y = 3\sin 2x$ and $y = 3 + \cos 2x$ on the same axes. Use your graph to solve the equation $3\sin 2x - \cos 2x = 3$.

11 Permutations and Combinations

11.1 Multiplication Principle

A student who drives from Home to School through the town centre can use either one of the two roads from his home to the town centre labelled 1 and 2, and then use any one of the three roads labelled A, B and C as shown in Figure 11.1.

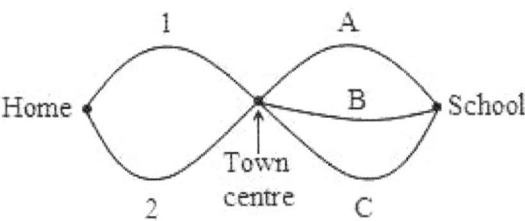

Figure 11.1

The different routes he can take are listed as follows: 1A, 1B, 1C, 2A, 2B, and 2C. In all, the number of possible routes is 6, which is simply the product of the number of routes from his home to the town centre with the number of routes from the town centre to the school.

This is an example of two general types of counting problems that can be solved using the Multiplication Principle: If there are m possible ways of making a first choice and for each of the m possible ways there are n possible ways of making a second choice then there are $m \times n$ possible ways of making the two choices.

Examples

1. How many different 4-digit numbers can be produced if the first digit may not be 0?

    We have 9 choices for the first digit, 10 choices for the second, third and fourth digits. Thus, the number of different numbers formed is

    $9 \cdot 10 \cdot 10 \cdot 10 = 9,000$

2. A class has 15 boys and 12 girls. In how many possible ways can one boy and one girl be selected to participate in a quiz?

    A boy can be selected in 15 different ways and a girl in 12 ways.

Thus, the number of possible selection is

$$15 \cdot 12 = 180$$

Exercise 11.1

1. A woman has 4 blouses and 5 skirts. How many different outfits can she wear?

2. A car company has 3 different car model and 6 colour schemes. How many cars must a dealer display to show each possibility?

3. A man has 3 pairs of shoes, 8 pairs of socks, 4 pairs of trousers and 9 shirts. How many outfits can he wear?

4. A school board has 12 members. The board must select from its members a chairman, vice-chairman and a secretary. In how many ways can this be done?

5. How many 4-letter code words are possible from the first 6 letters of the alphabet with no letters repeated?

6. How many ways are there to rank 7 candidates who apply for a job?

7. In how many ways can 3 boys and 3 girls be seated in a row of 6 seats if the seating must alternate boys and girls, starting with a girl?

8. How many 7-digit telephone numbers can be produced if the first digit may not be 0 and the digits are not repeated?

9. How many 7-digit telephone numbers can be assigned by a telephone company in a city if the first three digits are 446 and 447?

10. Using only the digits 0 and 1, how many different numbers consisting of 8 digits can be formed?

11. From a group of 5 people we are required to select a different person to participate in each of 3 different tests. In how many ways can the selections be made?

12. In how many ways can 10 blue flowers and 5 red flowers be planted in a row, if we do not distinguish between flowers of the same colour?

11.2 Permutation

Four boys A, B, C and D are to occupy four chairs in a row. Any one of the four boys can sit on the first chair:

$$A \quad B \quad C \quad D$$

Any one of the remaining three boys can sit on the second chair. The first two boys can be seated as listed below.

AB	AC	AD
BA	BC	BD
CA	CB	CD
DA	DB	DC

In all, there are 4 · 3 = 12 ways in seating the two boys. Any one of the two remaining boys can sit on the third chair. Therefore, the three boys can be seated in 4 · 3 · 2 = 24 ways. The remaining boy will sit on the fourth chair. Thus, the four boys can be seated in 4 · 3 · 2 · 1, written briefly as 4!, ways. The symbol 4! is read " 4 factorial ".

Generally, the number of ways n objects can be arranged in a particle order, denoted by ^nP_n, is $n!$. Each arrangement of a number of objects in a particular order is called permutation.

In arranging 5 objects, taken two objects at a time, we have five choices in selecting the first object and 4 choices in selecting the second object. Thus, the two objects can be arranged in 5 · 4 = 20 ways.

We can rewrite 5 · 4 as

$$\frac{5 \cdot 4 \cdot 3 \cdot 2 \cdot 1}{3 \cdot 2 \cdot 1} = \frac{5!}{3!}$$

$$= \frac{5!}{(5-2)!}$$

Therefore, the number of ways of arranging 5 objects taken 2 at a time, denoted by $^5P_2$ is given by

$$^5P_2 = \frac{5!}{(5-2)!} = 20$$

Generally, the number of permutations of n different objects taken r at a time, denoted as ^nP_r is given by

$$^nP_r = \frac{n!}{(n-r)!}$$

Examples

1. How many ways can you arrange the letters from the word FIGURE?

   Here we are arranging six letters using all of them. Thus, the number of ways the letters can be arranged is

   $^6P_6 = 6! = 720$

2. A class contains 12 students. In how many ways can prizes be awarded in 3 subjects, if no student can win more than one prize?

   We will have to choose 3 students from 12 students. The number of ways the prizes can be awarded is

   $$^{12}P_3 = \frac{12!}{(12-3)!}$$

   $$= \frac{12 \cdot 11 \cdot 10 \cdot 9!}{9!}$$

   $$= 1320$$

3. How many different four digit numbers can be formed using the five digits 0, 1, 2, 5 and 8, if no digit can be used more than once?

   Because the first digit cannot be 0, the first digit can be chosen in 4 ways. Whatever is chosen to be the first digit, the second digit can be chosen from 4 digits which include 0. Thus, the second digit can be chosen in 4 ways. The third digit can be chosen in 3 ways and the fourth digit in 2 ways. The number of ways the number can be formed is

   $$4 \cdot 4 \cdot 3 \cdot 2 = 96$$

Permutation with Conditions

Examples

1. Find number of arrangements of the letters in the word ABSENT that begin with a vowel.

There are two choices for the first letter as it must be filled by either A or E. To fill the second position we have a choice of five, one vowel together with four consonants. The third place can be filled in 4 ways, and so on. This gives $2 \cdot 5 \cdot 4 \cdot 3 \cdot 2 \cdot 1 = 240$ ways

2. A 4-digit even number greater than 2000 is formed by using four of the six digits 1, 2, 3, 5, 6 and 8. How many different numbers can be formed, if no digit may be used more than once in any number.

We will choose the last digit of the number first. Because the number is even it must end with the digit 2, 6 or 8. Thus, the last digit can be chosen in 3 ways. In order to make the number greater than 2000, the first digit cannot be 1, so there are 4 digits which can be chosen to be the first digit. When the first and last digits have been chosen, there are 4 digits left. The second digit can be chosen in 4 ways and the third digit in 3 ways. The number of four digit numbers formed is $4 \cdot 4 \cdot 3 \cdot 3 = 144$

Exercise 11.2

1. Ten athletes run in a race. Find the number of ways in which the first three places can be filled.

2. How many different ways are there to arrange the 6 letters in the word FRIDAY?

3. How many ways are there to rank 8 candidates who apply for a job?

4. A headmaster, a first assistant and a second assistant for a school are chosen from a group of 20 teachers. In how many ways can the 3 teachers be chosen for the 3 position?

5. A form captain and assistant captain for a class are to be chosen from a group of 10 students. In how many ways can the student be chosen for the two position?

6. From a pool of 10 job applicants, a list ranking the top 4 must be made. How many such list are possible?

7. A school bus has 16 seats. In how many different ways can 5 students be seated in it?

8. How many ways are there to arrange 7 people in a line?

9. How many ways can you arrange the letters from the word TYPES?

10. A telephone number contains 6 digits. How many possible telephone numbers are there in which no digit is repeated?

11. A school inspector must visit 6 schools. In how many different ways can he plan his visit?

12. In how many ways can the second runner-up, the first runner-up and the winner of a beauty pageant be chosen from 25 contestants?

13. The digits 0 through 9 are written on 10 cards. Four different cards are drawn and a 4-digit number is formed. How many different 4-digit numbers can be formed in this way?

14. A science class contains 30 students. In how many ways can prizes be awarded in mathematics, physics and chemistry, if no student can win more than one prize?

15. In how many ways can 6 different books be distributed to 8 children, if no child gets more than one book?

16. A computer must assign each of 5 output to one of 8 different printers. In how many ways can it do this provided no printer gets more than one output?

17. Find the number of three-figure integers that can be formed from the numbers 2, 3, 4, 5 and 6

    (a) if no number is used twice

    (b) if any number may be used more than once

18. How many numbers between 100 and 1000 can be formed which the digit 0 does not appear?

19. How many numbers between 5000 and 10, 000 can be formed, using only the digits 2, 3, 4, 6 and 8?

20. How many odd numbers between 500 and 1000 can be formed, using only the digits 1, 3, 6, 7 and 8?

21. How many five-figure odd numbers can be made from the digits 1, 2, 3, 4 and 5, if no digits is repeated?

22. How many even numbers greater than 2000 can be formed with the digits 1, 2, 4 and 8, if each digit may be used only once in each number?

23. How many numbers between 1000 and 10000 can be formed using only odd digits, if no digit is repeated?

24. Find the number of arrangements of the letters of the word MONDAY that begin either with A or O

25. Find the number of arrngements of the letters of the word AROUND in which the vowels and the consonants come alternately.

11.3 Combination

From the three letters A, B, C we choose two letters without repeating any letter. If order is important, we have the following selections.

$$AB \quad BA \quad AC \quad CA \quad BC \quad CB$$

This gives six different permutations of the three letters.

If we are concerned only with what combination of letters selected and not the particular order of the selection, only three selections can be made, namely: $\quad AB \quad AC \quad BC$

When a selection of objects is made without regard to order it is referred to as a combination. The number of combination of n objects taken r at a time is denoted by ^nC_r.

Because each selection consists of two letters that can be arranged in 2! ways, the number of permutation of the three letters taken two at a time is 2! times the number of combination of the three letters taken two at time. That is:

$$^3C_2 \times 2! = {^3P_2}$$

$$^3C_2 = \frac{^3P_2}{2!}$$

$$= \frac{3!}{2!(3-2)!}$$

$$= 3$$

CHAPTER 11 Permutations and Combinations

Generally, the number of combination of n objects taken r at a time is given by

$$^nC_r = \frac{n!}{r!(n-r)!}$$

Examples

1. A class is to select 3 students from 10 for a prize. In how many ways can the selection be made?

   The 3 students can be selected in $^{10}C_3$ ways.

   $$^{10}C_3 = \frac{10!}{3!(10-3)!}$$

   $$= \frac{10!}{3!7!}$$

   $$= 120$$

2. From eight men and seven women a committee of three is to be formed. In how many ways can this be done?

   $$^{15}C_3 = \frac{15!}{3!(15-3)!}$$

   $$= 455$$

3. A book shop has in stock six different mathematics books and five different science books. If two mathematics and three science books are to be selected, in how many ways can this be done?

   The two mathematics books can be chosen in $^6C_2$ ways. The science books can be chosen in $^5C_3$ ways. Now for each of the $^6C_2$ ways of selecting the mathematics books, there are $^5C_2$ ways of selecting the science books, therefore there are $^6C_2 \times {}^5C_2$ ways of selecting the books.

   $$^6C_2 \times {}^5C_2 = 15 \times 10 = 150$$

4. A mathematics test has two sections. There are five questions in Section A, and four questions in Section B. A student must answer three questions in all, with a least one question from each section. In how many different ways can a student complete the test?

A student can answer exactly two question from Section A and one question from Section B in $^5C_2 \times {}^4C_1$ ways, or one question from Section A and exactly two questions from Section B in $^5C_1 \times {}^4C_2$ ways. Thus, a student can complete the test in $^5C_2 \times {}^4C_1 + {}^5C_1 \times {}^4C_2 = 40 + 30 = 70$ ways.

Exercise 11.3

1. From eight students in a class a committee of three is to be formed. In how many ways can this be done?

2. In how many ways can a committee of 4 teachers be selected from a group of 9 teachers?

3. A school is to select 4 of 10 eligible students for a scholarship. In how many ways can the selection be made?

4. There are 25 members in a club. In how many ways can a subcommittee of 3 members be formed?

5. How many different relay teams of 4 persons can be chosen from a group of 12 runners?

6. A disciplinary committee has 3 teachers, 2 administration members and 3 students in it. In how many ways can a subcommittee of 2 teachers, 1 administrator and 1 student be formed?

7. A test has 2 parts. In Part 1 a student must do 3 of 5 questions and in Part 2 a student must choose 2 of 4 questions. In how many different ways can a student complete the test?

8. A sample of 8 persons is selected for a test from a group containing 10 students and 5 teachers. In how many ways can the 8 persons be selected?

9. A box contains 20 light builds. The quality control engineer will pick a sample of 4 light bulbs for inspection. How many different samples are there?

10. Nine points are marked on a circle. How many chords do they form?

11. There are 12 points on a plane, no three of them being in a straight line. How many triangles can be drawn using the points as vertices?

12. How many triangles can be formed by joining any three of the vertices of a heptagon?

13. Find the number of ways in which a group of eight boys can be split into two groups of four.

14. Three out of ten houses are to be painted white and the rest green. Find the number of ways in which the choice can be made.

15. A team of 11 players is to be chosen from 15 boys. Find the number of ways in which this can be done if there are two boys who refuse to play in the same team.

16. In how many ways can six girls be divided into two teams of three?

17. A school library contains 10 Mathematics and 8 Physics textbooks. In how many ways can a borrower select 2 of each?

18. A committee must contain 3 men and 4 women. In how many ways can the committee be chosen from 10 men and 6 women?

19. From 6 teachers and 5 students a committee of five is to be chosen that includes three students and two teachers. In how many ways can this be done?

20. How many committee of three can be formed from six girls and six boys, if the girls are in majority?

21. From 7 women and 5 men a committee of three is to be formed. The committee must include at least two women. In how many ways can this be done?

22. In how many ways can two taxis, each of which will take at most 4 passengers, take a party of 7 people, if 2 of them refuse to be in the same taxis?

23. There are five male teachers and seven female teachers. How many ways can they form a committee of four with a female as chairperson?

24. A team of 6 people is to be selected from 8 men and 5 women. Find the number of different teams that can be selected if:

   (a) there are no restriction

(b) the team contains 4 women

(c) the team contains at least 4 men

25. A committee of 7 members is to be selected from 6 teachers and 9 students. Find the number of different committees that may be selected if:

(a) there are no restriction.

(b) the committee consists of 2 teachers and 5 students.

(c) the committee must contain at most 1 teacher.

Review Exercise 11

1. How many different ways can the letters in the word STUDY be arranged?

2. Six athletes run in a race, assuming there are no ties, how many different results are possible?

3. How many ways can a captain and vice-captain be chosen from a football team of 11?

4. How many ways can three examination questions be chosen from seven?

5. If 6 couples go to a party, in how many ways can they pair off to dance?

6. Ama invited 4 friends to go with her to the school speech day. There are 120 different ways in which they can sit together in a row. In how many of those ways is Ama sitting in the middle?

7. Ten children are present at a party. In how many ways can four children be chosen to a game if two children refuse to play the game together?

8. How many 3-digit positive integers are odd and do not contain the digit 5?

9. From a box of 10 identical balls, 4 are to be removed. How many different sets of 4 balls could be removed?

10. A talent contest has 8 contestants. Judges must award prices for first, second and third places. If there are no ties, in how many different ways can the 3 prizes be awarded, and how many different groups of 3 people can get prizes?

170 CHAPTER 11   Permutations and Combinations

11. In how many ways can a student answer 8 out of 10 questions, if he must answer the first five questions?

12. How many 5-digits numbers can be formed from the nine digits, 1, 2, 3, ...,9

    (a) if no digit is repeated?

    (b) if repetitions are allowed?

13. (a) Find how many different 4-digit numbers can be formed from the digits 0, 3, 4, 5, 6 and 7 if each digit may be used only once.

    (b) Find how many of these 4-digit numbers are even.

14. How many different 4-digit numbers greater than 5000 can be formed using the digits 1, 3, 5, 6,7 and 8, if no digit can be used more than once?

15. A 4-digit number is formed by using four of the six digits 2, 3, 4, 5, 6 and 8, no digit may be used more than once in any number. How many different 4-digit numbers can be formed if:

    (a) there are no restriction.

    (b) the number is odd and more than 4000.

16. A committee of 5 members is to be selected from 6 women and 9 men. Find the number of different committees that may be selected if:

    (a) there are no restriction.

    (b) the committee must consist of 2 women and 3 men.

    (c) the committee must contain at least 3 women.

17. A committee of 6 members is to be selected from 8 teachers and 4 students. Find the number of different committees that can be selected if:

    (a) there are on restriction.

    (b) the committee contains all 4 students

    (c) the committee contains at least 4 teachers.

12 Binomial Expansions

12.1 The Binomial Theorem

Recall that $(a+b)^2 = a^2 + 2ab + b^2$. By the process of multiplication we can expand $(a+b)^3$ as follows.

$$(a+b)^3 = (a+b)(a+b)^2$$
$$= (a+b)(a^2 + 2ab + b^2)$$
$$= a^3 + 3a^2b + 3ab^2 + b^3$$

Using a similar process we get

$$(a+b)^4 = a^4 + 4a^3b + 6a^2b^2 + 4ab^3 + b^4$$

and

$$(a+b)^5 = a^5 + 5a^4b + 10a^3b^2 + 10a^2b^3 + 5ab^4 + b^5$$

Expanding $(a + b)$ raised to higher positive integral powers would be tedious and time consuming. The binomial theorem gives us an easy way to expand binomials.

You may have noticed the following.

1. In each expansion, there are $n + 1$ terms.

2. In each expansion, the power of a begin at n and decrease by 1, while the power of b begin with 0 and increase by 1.

3. The sum of the powers of each term is n.

4. The term in b^r is the $(r + 1)$ th term.

The coefficients of a binomial expansion are called binomial coefficients. The coefficient of $a^{n-r}b^r$ is given by

$$\binom{n}{r} = \frac{n!}{(n-r)!r!}$$

The notation ^nC_r is sometimes used in place of $\binom{n}{r}$.

The following Binomial Theorem gives you an easy way to expand binomials.

172 CHAPTER 12 Binomial Expansions

$$(a+b)^n = a^n + \binom{n}{1}a^{n-1}b + \binom{n}{2}a^{n-2}b^2 + \cdots + \binom{n}{r}a^{n-r}b^r + \cdots + n^n$$

Examples

1. Find each binomial coefficient.

   (a) $\binom{8}{3}$

   $$\binom{8}{3} = \frac{8!}{5!\cdot 3!} = \frac{8\cdot 7\cdot 6\cdot 5!}{5!\cdot 3!} = \frac{8\cdot 7\cdot 6}{3\cdot 2\cdot 1} = 56$$

   (b) $\binom{10}{4}$

   $$\binom{10}{4} = \frac{10!}{6!\cdot 4!} = \frac{10\cdot 9\cdot 8\cdot 7\cdot 6!}{6!\cdot 4!} = \frac{10\cdot 9\cdot 8\cdot 7}{4\cdot 3\cdot 2\cdot 1} = 210$$

2. Expand $(x-2y)^4$ using the binomial theorem.

   $$(x-2y)^4 = x^4 + \binom{4}{1}(x^3)(-2y) + \binom{4}{2}(x^2)(-2y)^2 + \binom{4}{3}(x)(-2y)^3 + (-2y)^4$$

   $$= x^4 - 8x^3y + 24x^2y^2 - 32xy^3 + 16y^4$$

3. Find, in ascending powers of x, the first five terms in the expansion of $(1+2x)^7$.

   $$(1+2x)^7 = 1 + \binom{7}{1}(2x) + \binom{7}{2}(2x)^2 + \binom{7}{3}(2x)^3 + \binom{7}{4}(2x)^4$$

   $$= 1 + 14x + 84x^2 + 280x^3 + 560x^4$$

4. Find the fourth term of $(x-2y)^9$.

   We use the fact that the $(r+1)$th term is given by $\binom{n}{r}a^{n-r}b^r$.

   Here $r + 1 = 4$, so $r = 3$. Because $n = 9$, $a = x$ and $b = -2y$, the fourth term in the binomial expansion is

   $$\binom{9}{3}(x^6)(-2y)^3 = 84 \cdot x^6 \cdot -8y^3$$

   $$= -672x^6y^3$$

The Binomial Theorem

5. Find the coefficient of x^4 in the expansion of $(2 + 3x)^7$.

   In this case, $n = 7$, $r = 4$, $a = 2$ and $b = 3x$. Substitute these values to obtain

   $$\binom{7}{4}(2)^3(3x)^4 = 22680x^4$$

   So, the coefficient is 22680.

Exercise 12.1

1. Evaluate the following binomial coefficients.

   (a) $\binom{9}{6}$ (b) $\binom{7}{0}$ (c) $\binom{10}{1}$

   (d) $\binom{6}{4}$ (e) $\binom{7}{3}$ (f) $\binom{10}{5}$

   (g) $\binom{12}{9}$ (h) $\binom{15}{8}$

2. Find the indicated term in the expansion of the following binomials.

   (a) $(1 + 4x)^6$, 3 rd term (b) $(3x - 2)^7$, 4 th term

   (c) $\left(1 - \frac{x}{2}\right)^{10}$, 6 th term (d) $(3 - 2x)^6$, 5 th term

3. Find the coefficient of the term indicated in the expansion of the binomials.

   (a) $(2x - y)^{12}$, term containing x^4

   (b) $\left(2x + \frac{1}{x}\right)^9$, term containing x^3

   (c) $(2x - 3y)^{10}$, term containing y^8

   (d) $\left(3x^2 - \frac{1}{2x}\right)^7$, term containing x^2

4. Expand the following expresssions.

   (a) $(x + y)^4$ (b) $(2x - y)^5$

   (c) $(3a + 2b)^4$ (d) $(4x - 3y)^3$

174 CHAPTER 12 Binomial Expansions

5. Find, in ascending powers of x, the first four terms in the expansion of the following binomials.

   (a) $(1-x)^{10}$ 　　　(b) $(1+2x)^9$ 　　　(c) $(2-x)^{10}$

   (d) $\left(1+\frac{1}{2}x\right)^6$ 　　　(e) $\left(3-\frac{2}{3}x\right)^7$ 　　　(f) $\left(2+\frac{1}{2}x\right)^8$

6. Find the term independent of x in the expansion of $\left(x^2-\frac{2}{x}\right)^6$.

7. Find the term in x in the expansion of $\left(x-\frac{2}{x}\right)^5$.

8. Find the middle term in the expansion of $\left(x-\frac{1}{y}\right)^8$.

9. Expand the following, up to the term indicated.

   (a) $(1+x-x^2)^7$, x^2 term

   (b) $(1-2x+3x^2)^6$, x^2 term

   (c) $(1-x)^7(1+x)$, x^3 term

   (d) $(1+2x)^5(1-x)$, x^3 term

10. Write down the first four terms of the binomial expansion of $(1-2x)^{10}$ and use it to find the value of $(0.98)^{10}$, correct to four decimal places.

11. Write down the first four terms of the binomial expansion of $(1+x/5)^6$ and use it to find the value of $(1.01)^6$, correct to four significant figures.

12. The first two terms of the expansion $(3+ax)^n$ are $243+810x$. Find the value of n and of a.

13. In the expansion of $(2+ax)^4$, where a is a positive integer, the coefficient of x^2 is equal to 2.25 times the coefficient of x. Find the value of a.

12.2 Pascal's Triangle

Consider the following expansions.

$(a + b)^0 = 1$

$(a + b)^1 = a + b$

$(a + b)^2 = a^2 + 2ab + b^2$

$(a + b)^3 = a^3 + 3a^2b + 3ab^2 + b^3$

$(a + b)^4 = a^4 + 4a^3b + 6a^2b^2 + 4ab^3 + b^4$

A triangular display of the numerical coefficient is shown below. This is called the Pascal's Triangle.

```
           1                    Zero row
         1   1                  First row
       1   2   1                Second row
     1   3   3   1              Third row
   1   4   6   4   1            Fourth row
```

The first and last numbers in each row of Pascal's Triangle are 1. Every other number in each row is formed by adding the two numbers immediately above the number. You can use the fourth row to find the fifth row as follows.

```
     1    4    6    4    1
      \  / \  / \  / \  /
   1    5   10   10   5    1
```

The n th row in Pascal's Triangle gives the coefficients of $(a + b)^n$.

Examples

1. Write the expansion of the following expressions.

(a) $(2x - y)^4$

Obtain the following binomial coefficients from the fourth row.

$$1, 4, 6, 4, 1$$

The expansion is worked out as follows.

$(2x - y)^4 = (2x)^4 + 4(2x)^3(-y) + 6(2x)^2(-y)^2 + 4(2x)(-y)^3 + (-y)^4$

$\qquad\quad = 16y^4 - 32x^3y + 24x^2y^2 - 8xy^3 + y^4$

(b) $(3x + 2y)^3$

Obtain the following binomial coefficients from the third row.

1, 3, 3, 1

So, the expansion is as follows.

$$(3x + 2y)^3 = (3x)^3 + 3(3x)^2(2y) + 3(3x)(2y)^2 + (2y)^3$$

$$= 27x^3 + 54x^2y + 36xy^2 + 8y^3$$

2. Expand $(3 + x)^4$ and use the first three terms to find 2.98^4.

$$(3 + x)^4 = (3)^4 + 4(3)^3(x) + 6(3)^2(x)^2 + 4(3)(x)^3 + x^4$$

$$= 81 + 108x + 54x^2 + 12x^3 + x^4$$

If x is made small successive terms may be so small that they would not affect the answer to the required degree of accuracy.

We rewrite 2.98^4 as $(3 - 0.02)^4 = [3 + (-0.02)]^4$

Replacing x with -0.02 in the expansion gives

$$2.98^4 = 81 + 108(-0.02) + 54(-0.02)^2$$

$$= 81 - 2.16 + 0.0216$$

$$= 78.8616$$

Note: When finding an approximation, take the first few terms of the expansion that will give you at least one place more than the required degree of accuracy.

Exercise 12.2

1. Expand the following expressions.

(a) $(1 + 2x)^6$ (b) $(x - 2y)^4$ (c) $(1 - x)^5$

(d) $(2x + 3y)^4$ (e) $(4x - 1)^3$ (f) $(3x + y)^4$

(g) $(1 - 3x)^4$ (h) $(2a + x)^6$ (i) $(2a - 3x)^4$

2. Expand $(1-x)^5$ and use the expansion to evaluate 0.98^5, correct to three significant figures.

3. Using the expansion of $(1+x)^4$ find the value of 1.01^4, correct to five significant figures.

4. Expand $(1-2x)^5$ and then use the expansion to evaluate 0.98^5, correct to three significant figures.

Review exercise 12

1. Find each binomial coefficient.

   (a) $\binom{12}{7}$ (b) $\binom{10}{6}$ (c) $\binom{15}{9}$

   (d) $\binom{16}{8}$ (e) $\binom{13}{5}$ (f) $\binom{25}{21}$

2. Use the Binomial Theorem to expand each expression.

   (a) $(x+2)^3$ (b) $(x+3)^5$ (c) $(x-y)^4$

   (d) $(2x-1)^5$ (e) $(2x+y)^6$ (f) $(1-2x)^4$

3. Use Pascal's Triangle to expand each expression.

   (a) $(x-3)^4$ (b) $(x+4)^3$ (c) $(2x-y)^3$

   (d) $(2x+y)^5$ (e) $(3x+2y)^4$ (f) $(4x-3y)^3$

4. Find, in ascending powers of x, the first four terms in the expansion of the following binomials.

   (a) $(2+x)^9$ (b) $(3-x)^7$ (c) $(1+2x)^{12}$

   (d) $(1-3x)^8$ (e) $\left(1+\tfrac{1}{2}x\right)^7$ (f) $\left(1-\tfrac{3}{2}x\right)^6$

5. Find the indicated term in the expansion of the following binomials.

   (a) $(x-y)^{10}$, 4 th term   (b) $(3x+y)^{12}$, 10 th term

   (c) $(3x+y)^7$, 3 rd term   (d) $(a-4b)^9$, 6 th term

6. Find the coefficient of the term indicated in the expansion of each binomial.

   (a) $(x-1)^{10}$, term containing x^7 (b) $(x+y)^{15}$, term containing y^{11}

   (c) $(2x-y)^{12}$, term containing y^9 (d) $(x+y)^{10}$, term containing y^3

   (e) $(x^2+3)^4$, term containing x^4 (f) $(x-3)^{12}$, term containing x^9

7. Find the term independent of x in the expansion of the following expressions.

   (a) $\left(x^2 - \dfrac{2}{x}\right)^6$ (b) $\left(x^3 + \dfrac{1}{2x}\right)^8$ (c) $\left(2x - \dfrac{1}{x^4}\right)^{10}$

8. Write down the binomial expansion of $(x+y)^4$. Use your expansion to evaluate, correct to three decimal places 1.99^4.

9. (a) Using the Binomial Theorem, write down the expansion of $(1+x)^7$.

   (b) Use your result in (a) to evaluate 0.997^7, correct to five decimal places.

10. (a) Write down and simplify the first five terms of the binomial expansion of $(1 - \tfrac{1}{3}x)^8$

    (b) Use your expansion to evaluate 0.997^8, correct to five decimal places.

11. Expand $(1-2x)^6$ and then use the expansion to evaluate 0.94^6, correct to three significant figures.

12. The coefficient of x^2 in the expansion of $(2+px)^6$ is 15. Find the value of p.

13. The coefficient of x^2 in the expansion of $(1+px)^5$ is 90. Find the value of p.

14. Expand the following, up to the term in x^3.

    (a) $(3-x)\left(2+\tfrac{1}{4}x\right)^6$ (b) $\left(1+\tfrac{1}{2}x\right)(1-2x)^6$

    (c) $(1+2x)(2-x)^8$ (d) $(2+x)^6(1-x)^2$

13 Vectors

13.1 Vector Algebra

Quantities such as displacement, force and velocity have both magnitude (size or length) and direction. Such quantities are called vectors. Some quantities have magnitude only. These quantities are called scalars. Examples include time, temperature, mass and volume.

Vectors are represented geometrically by a directed line segment, as shown in Figure 13.1.

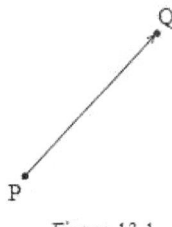

Figure 13.1

The directed line segment which has initial point P and terminal point Q represents a vector, denoted by $\overrightarrow{PQ}$.

The magnitude of the vector is denoted by $|\overrightarrow{PQ}|$ or PQ

In print vectors are denoted by boldface letters such as **AB** or **a**. In handwritting a vector can be denoted by an underlined letter such as $\underline{a}$ or a letter with an arrow above such as $\vec{a}$ or $\vec{A}$

Vectors Written in Component Form

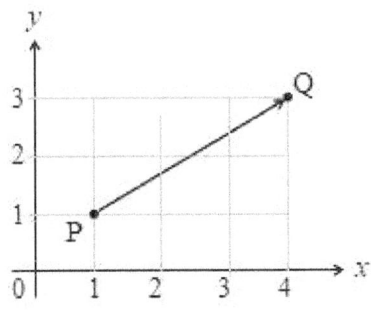

Figure 13.2

180   CHAPTER 13   Vectors

The vector $\overrightarrow{PQ}$ shown in Figure 13.2 can be written in terms of its movement in the *x*- and *y*- directions as $\overrightarrow{PQ} = \begin{pmatrix} 3 \\ 2 \end{pmatrix}$. The vector is written in component form.

The movement in the *x*-direction is called the horizontal component, and the movement in the *y*- direction is called the vertical component. In the *x*-direction movement to the right is considered positive and movement to left is negative. In the *y*-direction movement up is considered positive and movement down is negative.

Equal Vectors

Two or more vectors that have the same magnitude and direction are said to be equal. If $\overrightarrow{AB}$ and $\overrightarrow{PQ}$ are equal vectors, we write $\overrightarrow{AB} = \overrightarrow{PQ}$.

Addition of Vectors

Two vectors can be added geometrically by using the following laws.

The Triangle Law

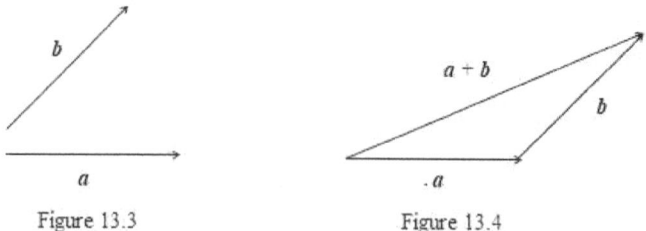

Figure 13.3   Figure 13.4

To add vector *a* and vector *b* shown in Figure 13.3 place the initial point of *b* on the terminal point of *a*, as shown in Figure 13.4. The sum *a* + *b* of the vectors is formed by joining the initial point of *a* with the terminal point of *b*. The sum *a* + *b* is called the resultant of vector addition of *a* and *b*.

The Parallelogram Law

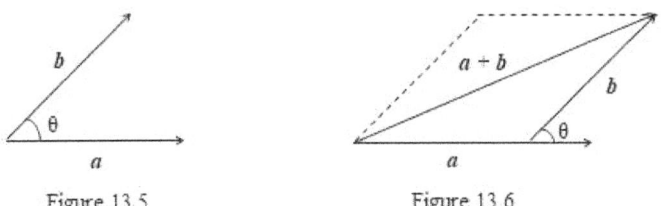

Figure 13.5   Figure 13.6

To add the vectors shown in Figure 13.5 we translate the initial point of vector b to the terminal point of vector a, and then complete the parallelogram as shown in Figure 13.6. The resultant of vector addition of a and b is the diagonal of the parallelogram having a and b as its adjacent sides.

Adding Vectors in Component Form

To find the sum of two vectors in component form, we add the corresponding components. If

$$a = \begin{pmatrix} a_1 \\ a_2 \end{pmatrix} \quad \text{and} \quad b = \begin{pmatrix} b_1 \\ b_2 \end{pmatrix},$$

then the sum a and b is the vector

$$a + b = \begin{pmatrix} a_1 + b_1 \\ a_2 + b_2 \end{pmatrix}$$

Examples

Find the following vectors.

1. $\begin{pmatrix} -3 \\ 2 \end{pmatrix} + \begin{pmatrix} 5 \\ -3 \end{pmatrix}$

$$\begin{pmatrix} -3 \\ 2 \end{pmatrix} + \begin{pmatrix} 5 \\ -3 \end{pmatrix} = \begin{pmatrix} 2 \\ -1 \end{pmatrix}$$

2. $\begin{pmatrix} 2 \\ -3 \end{pmatrix} + \begin{pmatrix} -5 \\ 1 \end{pmatrix}$

$$\begin{pmatrix} 2 \\ -3 \end{pmatrix} + \begin{pmatrix} -5 \\ 1 \end{pmatrix} = \begin{pmatrix} -3 \\ -2 \end{pmatrix}$$

CHAPTER 13 Vectors

Exercise 13.1(a)

1. Find:

   (a) $\binom{2}{3} + \binom{1}{-4}$ (b) $\binom{3}{-1} + \binom{-4}{3}$ (c) $\binom{3}{-4} + \binom{-3}{5}$

   (d) $\binom{2}{4} + \binom{-3}{-5}$ (e) $\binom{-7}{2} + \binom{1}{5}$ (f) $\binom{-2}{-3} + \binom{5}{1}$

2. Find the value of x and of y from the following vector equations.

   (a) $\binom{3}{x} + \binom{y}{2} = \binom{8}{7}$

   (b) $\binom{1}{2} + \binom{y}{-5} = \binom{3}{x}$

   (c) $\binom{2x+3}{x} + \binom{-3}{3y-2} = \binom{y}{5}$

   (d) $\binom{3x}{2x-y} + \binom{y}{-7} = \binom{2-3y}{2y}$

3. Given that

   $a = \binom{2}{-1}, b = \binom{x}{y}, c = \binom{-1}{3}$ and $a + b = c$.

   Find the value of x and of y.

Negative Vectors

A vector b that has the same magnitude as a but opposite in direction, as shown in Figure 13.7, is called the negative vector of a, and we write $b = -a$.

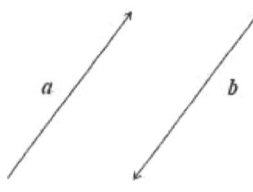

Figure 13.7

If $a = \binom{a_1}{a_2}$, then the negative vector of a is $\binom{-a_1}{-a_2}$.

Vector Algebra

Example

Find $-\begin{pmatrix} -3 \\ 4 \end{pmatrix}$

$$-\begin{pmatrix} -3 \\ 4 \end{pmatrix} = \begin{pmatrix} 3 \\ -4 \end{pmatrix}$$

Exercise 13.1(b)

1. Find:

    (a) $-\begin{pmatrix} 1 \\ 3 \end{pmatrix}$ \qquad (b) $-\begin{pmatrix} 3 \\ -4 \end{pmatrix}$

    (c) $-\begin{pmatrix} -2 \\ 5 \end{pmatrix}$ \qquad (d) $-\begin{pmatrix} -2 \\ 0 \end{pmatrix}$

2. Given that the vectors $a = \begin{pmatrix} 2 \\ 3 \end{pmatrix}, b = \begin{pmatrix} -2 \\ 4 \end{pmatrix}$ and $c = \begin{pmatrix} 4 \\ 1 \end{pmatrix}$, find:

    (a) $-a$ \qquad (b) $-b$

    (c) $-c + b$ \qquad (d) $-b + a$

Subtraction of Vectors

If a and b are two vectors the difference vector $a - b$ is equal to $a + (-b)$.

Example

Find $\begin{pmatrix} -3 \\ 2 \end{pmatrix} - \begin{pmatrix} -5 \\ 6 \end{pmatrix}$

$$\begin{pmatrix} -3 \\ 2 \end{pmatrix} - \begin{pmatrix} -5 \\ 6 \end{pmatrix} = \begin{pmatrix} 2 \\ -4 \end{pmatrix}$$

Exercise 13.1(c)

1. Find:

    (a) $\begin{pmatrix} 3 \\ 5 \end{pmatrix} - \begin{pmatrix} 1 \\ 2 \end{pmatrix}$ \qquad (b) $\begin{pmatrix} 5 \\ 2 \end{pmatrix} - \begin{pmatrix} -2 \\ 3 \end{pmatrix}$

(c) $\begin{pmatrix}1\\2\end{pmatrix} - \begin{pmatrix}1\\-2\end{pmatrix}$ (d) $\begin{pmatrix}-7\\-6\end{pmatrix} - \begin{pmatrix}-5\\3\end{pmatrix}$

(e) $\begin{pmatrix}-2\\1\end{pmatrix} - \begin{pmatrix}1\\2\end{pmatrix}$ (f) $\begin{pmatrix}-2\\1\end{pmatrix} - \begin{pmatrix}-8\\-4\end{pmatrix}$

2. Given that $a = \begin{pmatrix}2\\-3\end{pmatrix}$, $b = \begin{pmatrix}3\\-4\end{pmatrix}$ and $c = \begin{pmatrix}4\\1\end{pmatrix}$, find:

(a) $b - a$ (b) $c - a$ (c) $b - c$

Multiplying Vectors by Scalars

The product of a vector a and a scalar k is the vector that is $|k|$ times as long as a. If k is positive, ka has the same direction as a, and if k is negative, ka has the direction opposite of a, as shown in Figure 13.8.

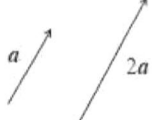 

Figure 13.8

To multiply the vector $a = \begin{pmatrix}a_1\\a_2\end{pmatrix}$ by a scalar k multiply both components by k. That is $ka = \begin{pmatrix}ka_1\\ka_2\end{pmatrix}$.

Examples

1. Find $-3\begin{pmatrix}1\\-2\end{pmatrix}$

$$-3\begin{pmatrix}1\\-2\end{pmatrix} = \begin{pmatrix}-3\\6\end{pmatrix}$$

2. Given $a = \begin{pmatrix}3\\-2\end{pmatrix}$ and $b = \begin{pmatrix}-1\\4\end{pmatrix}$, find:

(a) $2a + 3b$

$$2a + 3b = 2\begin{pmatrix}3\\-2\end{pmatrix} + 3\begin{pmatrix}-1\\4\end{pmatrix}$$

$$= \begin{pmatrix}6\\-4\end{pmatrix} + \begin{pmatrix}-3\\12\end{pmatrix}$$

$$= \begin{pmatrix}3\\8\end{pmatrix}$$

(b) $3a - b$

$$3a - b = 3\begin{pmatrix}3\\-2\end{pmatrix} - \begin{pmatrix}-1\\4\end{pmatrix}$$

$$= \begin{pmatrix}9\\-6\end{pmatrix} - \begin{pmatrix}-1\\4\end{pmatrix}$$

$$= \begin{pmatrix}10\\-10\end{pmatrix}$$

Exercise 13.1(d)

1. Find:

   (a) $2\begin{pmatrix}3\\-1\end{pmatrix}$ (b) $-3\begin{pmatrix}-1\\2\end{pmatrix}$ (c) $\frac{1}{2}\begin{pmatrix}-2\\6\end{pmatrix}$

   (d) $-\frac{2}{3}\begin{pmatrix}3\\-6\end{pmatrix}$ (e) $-2\begin{pmatrix}-3\\0\end{pmatrix}$ (f) $\frac{4}{3}\begin{pmatrix}-3\\-6\end{pmatrix}$

2. Given that $a = \begin{pmatrix}1\\2\end{pmatrix}, b = \begin{pmatrix}2\\-3\end{pmatrix}$ and $c = \begin{pmatrix}3\\4\end{pmatrix}$, find:

   (a) $-2a$ (b) $3b$ (c) $3a + 2b$ (d) $2b - c$

3. Given that $a = \begin{pmatrix}2\\3\end{pmatrix}$ and $b = \begin{pmatrix}3\\-1\end{pmatrix}$, solve the equation
   $xa + yb = \begin{pmatrix}12\\7\end{pmatrix}$.

4. Given that $a = \begin{pmatrix}3\\4\end{pmatrix}, b = \begin{pmatrix}2\\-3\end{pmatrix}$ and $c = \begin{pmatrix}5\\1\end{pmatrix}$, solve the equation
   $xa + yb = c$.

5. Given that $a = \begin{pmatrix}3\\2\end{pmatrix}, b = \begin{pmatrix}2\\-3\end{pmatrix}$ and $c = \begin{pmatrix}12\\-5\end{pmatrix}$, find k and m such that
   $ka + mb = c$.

Magnitude of Vectors

The magnitude (or length) of a vector is represented geometrically by the length of a line segment.

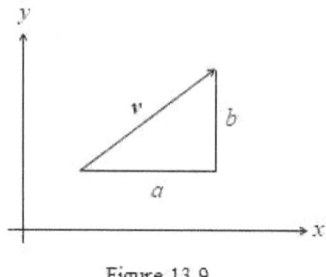

Figure 13.9

Consider the vector $v = \begin{pmatrix} a \\ b \end{pmatrix}$ shown in Figure 13.9. The Pythagoras' Theorem gives the magnitude of the vector v as

$$|v| = \sqrt{a^2 + b^2}$$

Examples

1. Find the magnitude of the vector $a = \begin{pmatrix} -8 \\ 15 \end{pmatrix}$.

$$|v| = \sqrt{(-8)^2 + 15^2}$$

$$= \sqrt{289}$$

$$= 17 \text{ units}$$

2. Given that $a = \begin{pmatrix} -3 \\ 2 \end{pmatrix}$ and $b = \begin{pmatrix} 3 \\ -8 \end{pmatrix}$, find $|2a + b|$.

$$2a + b = 2\begin{pmatrix} -3 \\ 2 \end{pmatrix} + \begin{pmatrix} 3 \\ -8 \end{pmatrix}$$

$$= \begin{pmatrix} -6 \\ 4 \end{pmatrix} + \begin{pmatrix} 3 \\ -8 \end{pmatrix}$$

$$= \begin{pmatrix} -3 \\ -4 \end{pmatrix}$$

$$|2a + b| = \sqrt{(-3)^2 + (-4)^2}$$
$$= \sqrt{25}$$
$$= 5 \text{ units}$$

Exercise 13.1(e)

1. Find the magnitude of each of the following vectors.

   (a) $\begin{pmatrix} 4 \\ 3 \end{pmatrix}$ (b) $\begin{pmatrix} -2 \\ 4 \end{pmatrix}$

   (c) $\begin{pmatrix} -3 \\ 5 \end{pmatrix}$ (d) $\begin{pmatrix} -5 \\ -12 \end{pmatrix}$

2. Given that $a = \begin{pmatrix} -7 \\ 2 \end{pmatrix}$ and $b = \begin{pmatrix} 1 \\ 5 \end{pmatrix}$, evaluate :

   (a) $|a + b|$ (b) $|b - a|$

   (c) $|2a + b|$ (d) $|a - 2b|$

Unit Vectors

A vector that has a magnitude of 1 is called a unit vector. The unit vector that has the same direction as a vector a is given by

$$\text{Unit vector} = \frac{a}{|a|}$$

Example

Find the unit vector in the direction of $a = \begin{pmatrix} 3 \\ -4 \end{pmatrix}$.

$$|a| = \sqrt{3^2 + (-4)^2}$$
$$= \sqrt{25}$$
$$= 5 \text{ units}$$

The unit vector of a is

$$\frac{1}{5}\begin{pmatrix} 3 \\ -4 \end{pmatrix} = \begin{pmatrix} 3/5 \\ -4/5 \end{pmatrix}.$$

Standard Unit Vectors

Two units vectors, one in the x-direction, denoted by *i* and the other in the y-direction, denoted by *j* are called standard unit vectors. Note that

$$i = \begin{pmatrix} 1 \\ 0 \end{pmatrix} \text{ and } j = \begin{pmatrix} 0 \\ 1 \end{pmatrix}.$$

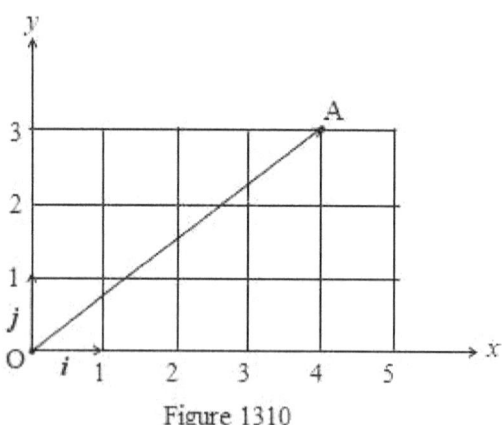

Figure 1310

The vector $\overrightarrow{OA}$ shown in Figure 13.10 can be written as

$$\overrightarrow{OA} = 4i + 3j$$

The vector sum $4i + 3j$ is called a linear combination of the vectors *i* and *j*. In general, the vector $a = \begin{pmatrix} a_1 \\ a_2 \end{pmatrix}$ can be written as a linear combination of the standard unit vectors *i* and *j* as $a = a_1 i + a_2 j$. The scalars a_1 and a_2 are called the horizontal and vertical components of *a* respectively.

Examples

1. Given that $a = 3i + 2j$ and $b = -5i + j$, find:

   (a) $a + b$

$$a + b = (3i + 2j) + (-5i + j)$$
$$= (3 - 5)i + (2 + 1)j$$
$$= -2i + 3j$$

(b) $2a - b$

$$2a - b = 2(3i + 2j) - (-5i + j)$$
$$= (6i + 4j) - (-5i + j)$$
$$= (6 + 5)i + (4 - 1)j$$
$$= 11i + 3j$$

2. Given that $a = 3i - 4j$, find $|-3a|$.

$$-3a = -3(3i - 4j)$$
$$= -9i + 12j$$
$$|-3a| = \sqrt{(-9)^2 + 12^2}$$
$$= \sqrt{225}$$
$$= 15 \text{ units}$$

3. Find the unit vector in the direction of $a = -5i + 12j$.

$$|a| = \sqrt{(-5)^2 + 12^2},$$
$$= \sqrt{169}$$
$$= 13 \text{ units}$$

The unit vector is

$$-\frac{5}{13}i + \frac{12}{13}j.$$

Exercise 13.1(f)

1. Express each vector in terms of i and j.

(a) $\begin{pmatrix} 4 \\ 3 \end{pmatrix}$ (b) $\begin{pmatrix} 3 \\ -2 \end{pmatrix}$ (c) $\begin{pmatrix} -5 \\ -2 \end{pmatrix}$ (d) $\begin{pmatrix} 0 \\ 5 \end{pmatrix}$ (e) $\begin{pmatrix} -7 \\ 0 \end{pmatrix}$

2. Given that $a = -3i + 2j$, $b = 2i - 3j$ and $c = 5i + 4j$, find:

(a) $a + b$ (b) $b - c$ (c) $2c + b$ (d) $3a - 2b$

190 CHAPTER 13 Vectors

3. Find the magnitude of each vector.

   (a) $-3i + j$ (b) $4i - 3j$ (c) $-12j$ (d) $-15i - 8j$

4. Find the unit vector in the direction of the following vectors.

   (a) $i + j$ (b) $-3i + 2j$ (c) $2i - j$ (d) $-i - 7j$

Position Vectors

The vector from the origin to a point is the position vector of that point. The position vector of a point is represented uniquely by the coordinates of its terminal point.

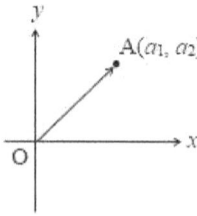

Figure 13.11   Figure 13.12

Figure 13.11 shows a vector which has its initial point at the origin O and its terminal point at the point A, having coordinates (a_1, a_2). The position vector of A, denoted by $\overrightarrow{OA}$ or a is given by $a = \begin{pmatrix} a_1 \\ a_2 \end{pmatrix}$.

In Figure 13.12, the vector $\overrightarrow{AB}$ has intial point $A(a_1, a_2)$ and terminal point $B(b_1, b_2)$.

$$\overrightarrow{AB} = \overrightarrow{AO} + \overrightarrow{OB}$$
$$= \overrightarrow{OB} - \overrightarrow{OA}$$
$$= \begin{pmatrix} b_1 \\ b_2 \end{pmatrix} - \begin{pmatrix} a_1 \\ a_2 \end{pmatrix}$$
$$= \begin{pmatrix} b_1 - a_1 \\ b_2 - a_2 \end{pmatrix}$$

Examples

1. Find the vector $\vec{AB}$ that has initial point A(− 7, 5) and terminal point B(− 2, 3).

$$\vec{AB} = \vec{OB} - \vec{OA}$$

$$= \begin{pmatrix} -2 \\ 3 \end{pmatrix} - \begin{pmatrix} -7 \\ 5 \end{pmatrix}$$

$$= \begin{pmatrix} 5 \\ -2 \end{pmatrix}$$

2. A vector a has initial point (3, − 6) and terminal point (− 2, 6). Find $|a|$.

$$a = \begin{pmatrix} -2 \\ 6 \end{pmatrix} - \begin{pmatrix} 3 \\ -6 \end{pmatrix}$$

$$= \begin{pmatrix} -5 \\ 12 \end{pmatrix}$$

$$|a| = \sqrt{(-5)^2 + 12^2}$$

$$= \sqrt{169}$$

$$= 13 \text{ units}$$

3. Given that P(− 3, 2) and $\vec{PQ} = \begin{pmatrix} 5 \\ 3 \end{pmatrix}$, find the coordinates of Q.

$$\vec{PQ} = \vec{OQ} - \vec{OP}$$

$$\begin{pmatrix} 5 \\ 3 \end{pmatrix} = \vec{OQ} - \begin{pmatrix} -3 \\ 2 \end{pmatrix}$$

$$\vec{OQ} = \begin{pmatrix} 5 \\ 3 \end{pmatrix} + \begin{pmatrix} -3 \\ 2 \end{pmatrix}$$

$$= \begin{pmatrix} 2 \\ 5 \end{pmatrix}$$

The coordinates of Q is (2, 5).

Exercise 13.1(g)

1. Write the position vectors of the following points.

   (a) (4, 2)   (b) (−5, 3)   (c) (0, 4)   (d) (−3, −2)   (e) (−5, 0)

2. Write the position vectors of the following points in terms of i and j.

   (a) (3, 2)   (b) (1, 3)   (c) (−2, 0)   (d) (4, −5)   (e) (0, 2)

3. Find the vector $\overrightarrow{AB}$ in each case.

   (a) A(7, 3), B(9, 4)   (b) A(1, −2), B(3, −5)

   (c) A(3, 4), B(2, −1)   (d) A(2, 3), B(−2, 4)

4. Given A(x, y), B(3, 4) and $\overrightarrow{AB} = \begin{pmatrix} -2 \\ 1 \end{pmatrix}$, find the value of x and of y.

5. Given A(4, −5) and $\overrightarrow{AB} = \begin{pmatrix} 2 \\ 3 \end{pmatrix}$, find the coordinates of B.

6. Given Q(−2, −3) and $\overrightarrow{PQ} = \begin{pmatrix} -3 \\ 4 \end{pmatrix}$, find the coordinates of P.

7. The position vectors relative to the origin O, of three points A, B and C are $2i + 3j$, $4i + 7j$ and $i + 2j$ respectively. Given that $\overrightarrow{OB} = m\overrightarrow{OA} + n\overrightarrow{OC}$, where m and n are scalar constants, find the value of m and of n.

8. The position vectors of A and B are $-3i + 2j$ and $-i - j$. Given that $\overrightarrow{AC} = 5i + j$, find $\overrightarrow{BC}$.

9. A and B have position vectors $5i + 2j$ and $-i + 4j$ respectively. Find the position vector of C if $3\overrightarrow{OA} = 2\overrightarrow{OB} + \overrightarrow{OC}$.

10. The coordinates of A and B are (2, 3) and (−2, 5) respectively. Find the position vector of C if $2\overrightarrow{OA} = 2\overrightarrow{OB} + \overrightarrow{BC}$.

13.2 Composition and Resolution of Velocities

Composition of Velocities

Two or more velocites can be composed into a single resultant velocity which will represent them.

In real life, some objects move within a medium that is moving with respect to an observer. Examples are a boat moving in a river and an airplane flying in the air.

Suppose that an airplane flying from A to B with a velocity *a* encounters a wind blowing from the west with velocity *b*, as illustrated in Figure 13.13.

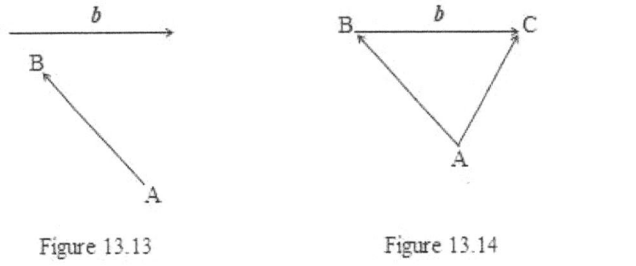

Figure 13.13   Figure 13.14

Due to the wind an observer on the ground will see the plane flying along the line AC, as shown in Figure 13.14. The vector $\overrightarrow{AC}$, which is the resultant vector, represents the velocity relative to the ground, called the ground speed, and the direction of the airplane as seen from the ground, called the track.

Notice that the velocity of the airplane relative to the ground is the vector sum of the velocity of the airplane relative to the air, called the airspeed and the velocity of the wind relative to the ground, called the wind speed. The direction in which the pilot steers an airplane is called its course.

The effect of the river current upon a boat is similar to the effect of the wind upon an airplane. The velocity of a boat as seen by an observer on the ground is the vector sum of the velocity of the boat in the river and the velocity of the current.

Resolution of Velocites

A single velocity can be split into two components which are at right angles to each other.

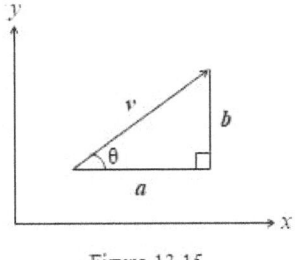

Figure 13.15

Figure 13.15 shows a velocity vector v which makes angle θ with the horizontal. The vector is resolved into two perpendicular components a and b along the x- direction and the y-direction respectively. By trigonometry the component along the x-direction is $a = v \cos \theta$ and that along the y-direction is $b = v \sin \theta$, so v can be written as

$$v = \begin{pmatrix} v \cos \theta \\ v \sin \theta \end{pmatrix}$$

or $v = v \cos \theta \, i + v \sin \theta \, j$

The magnitude of v is given by

$$|v| = \sqrt{a^2 + b^2}.$$

Notice that $\tan \theta = (b/a)$. It follows that, the angle which the vector v makes with the horizontal is $\tan^{-1}(b/a)$.

The resultant of two or more velocities can be obtained by resolving each velocity and then adding the corresponding components.

Examples

1. An airplane is flying on a bearing of 050° with an airspeed of 120 km h^{-1}. There is a wind of 35 km h^{-1} from north-west. Find the ground speed and the track of the plane.

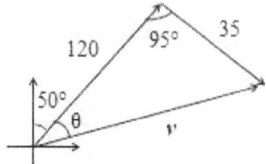

Let v represents the velocity of the airplane relative to the ground.

The velocity of the airplane is $\begin{pmatrix} 120 \cos 40° \\ 120 \sin 40° \end{pmatrix}$

and the velocity of the wind is $\begin{pmatrix} 35 \cos 45° \\ -35 \sin 45° \end{pmatrix}$.

$$v = \begin{pmatrix} 120 \cos 40° \\ 120 \sin 40° \end{pmatrix} + \begin{pmatrix} 35 \cos 45° \\ -35 \sin 45° \end{pmatrix}$$

$$= \binom{116.67}{52.39}$$

$$|v| = \sqrt{116.67^2 + 52.39^2}$$

$$= 127.9$$

The ground speed is 127.9 km h^{-1}.

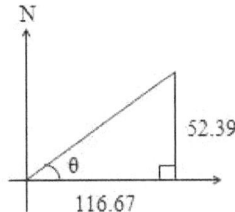

From the diagram, we get

$$\tan \theta = \frac{52.39}{116.67}$$

$$\theta = 24.2°$$

The track of the plane is 065.8°.

You can obtain the same result using trigonometry as follows.

$$|v|^2 = 120^2 + 35^2 - 2(120)(35) \cos 95°$$

$$= 16357.11$$

$$|v| = \sqrt{16357.11}$$

$$= 127.9 \text{ km h}^{-1}$$

$$\frac{\sin \theta}{35} = \frac{\sin 95°}{127.9}$$

$$\sin \theta = \frac{35 \sin 95°}{127.9}$$

$$\theta = 15.8°$$

The track is 065.8°.

2. A pilot flies his airplane directly from a point A to a point B, distance of 300 km. The bearing of B from A is 315°. A wind of 50 km h^{-1} is blowing from the west. Given the airplane can travel at 164 km h^{-1} in still air, find:

(a) the bearing on which the plane must be steered.

(b) the time taken to fly from A to B.

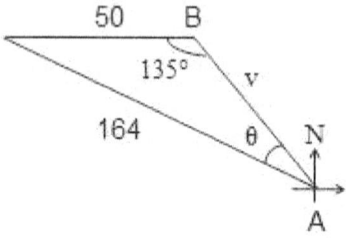

$$\frac{\sin \theta}{50} = \frac{\sin 135°}{164}$$

$$\sin \theta = \frac{50 \sin 135°}{164}$$

$$\theta = 12.4°$$

The bearing is 302.6°

$$\frac{v}{\sin 32.6°} = \frac{164}{\sin 135°}$$

$$v = \frac{164 \sin 32.6°}{\sin 135°}$$

$$= 125 \text{ km h}^{-1}$$

$$\text{Time} = \frac{300}{125}$$

$$= 2.4 \text{ hours}$$

3. A river with parallel banks flows to the north. The river is 120 metres wide. A motor boat travelling with a speed of 8 m s^{-1} encounters a current flowing with speed 5 m s^{-1}.

(a) What is the velocity of the boat relative to an observer on the shore.

(b) If the boat travels in a straight line from a point P on one bank to a point Q which is on the other bank directly opposite P, find the angle to the bank at which the boat should be steered.

(a)

Since the boat heads straight across the river and since the current is always directed straight up, the two vectors are at right angles to each other. Hence, we can use the Pythagoras' theorem to determine the resultant velocity.

Let v be the velocity of the boat relative to an observer.

$$|v|^2 = 8^2 + 5^2$$
$$= 89$$
$$|v| = \sqrt{89}$$
$$= 9.4 \text{ m s}^{-1}$$

(b)

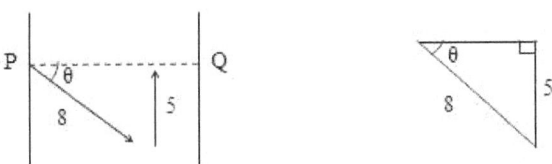

Let θ be the angle to the bank.

$$\sin \theta = \frac{5}{8}$$
$$\theta = 38.7°$$

The boat should travel at 38.7° to the line PQ directly across the river.

Exercise 13.2

1. A pilot is steering an aircraft due east and its airspeed is 600 km h^{-1}. There is a wind blowing from the south of 50 km h^{-1}. Find the direction in which the aircraft travel and its speed over the ground.

2. A pilot was steering his aircraft on a bearing of 136° and his airspeed indicator showed 200 km h^{-1}. However, there was a wind blowing at 50 km h^{-1} from the west. What was its speed over the ground and direction.

3. An airplane whose ground speed is 450 km h^{-1} travels on a bearing of 300°. There is a wind blowing from north-east of 80 km h^{-1}. Find the airspeed and the course of the airplane.

4. An airplane is flying in the direction of 145° with airspeed of 800 km h^{-1}. Because of the wind, its ground speed and direction are 750 km h^{-1} and 130° respectively. Find the direction and speed of the wind.

5. A helicopter leaves an airfield A to fly to B 500 km away on a bearing of 140°. There is a steady wind of 30 km h^{-1} from north-east. The helicopter has airspeed of 150 km h^{-1}. Find the course the pilot must take and the time taken for him to reach B.

6. An airplane whose speed in still air is 450 km h^{-1} travels directly from A to B, distance of 1200 km. The bearing of B from A is 215° and there is a wind of 60 km h^{-1} from the east.

    (a) Find the bearing on which the plane was steered.

    (b) Find the time taken for the journey.

7. A river is flowing at 3 m s^{-1}. A man can row at 5 m s^{-1}. Find the angle to the bank that he should row if he is to cross the river directly. If the river is 200 m wide how long will it take him?

8. A river is flowing at 2 m s^{-1}. A man can row at 2.5 m s^{-1}. In which direction should he row if he is to cross the river directly? If the river is 60 m wide how long will it take him?

9. A ship sets off on a bearing of 147° at a speed of 30 km h^{-1} through the water. The current flows at 5 km h^{-1} in the direction 083°. Find the direction the ship travels and the distance of the ship after 3 hours?

10. A river with parallel banks is 50 m wide and is flowing east at 1.2 m s^{-1}. A point Q on the other bank is directly opposite a point P. If a boat takes 10 seconds to travel in a straight line from P to Q, calculate the speed of the boat in still water and the angle to the bank at which the boat should be steered.

13.3 Relative Velocity

The velocity of one body relative to another is called relative velocity. For example, the velocity of bus B as seen by an observer in car C is the relative velocity of B to C, written V_{BC} (or $_BV_C$). The relative velocity of B to C is given by the equation $V_{BC} = V_B - V_C$

Example

A particle A moves east at 25 m s^{-1}, and a particle B moves north at 40 m s^{-1}. Find the velocity of B relative to A.

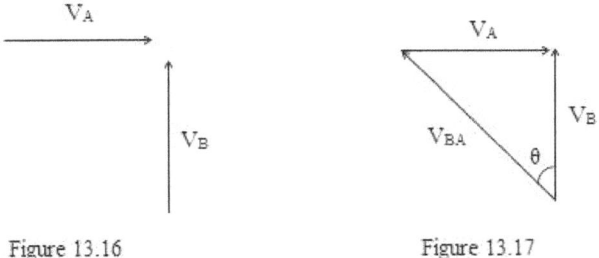

Figure 13.16   Figure 13.17

The relative velocity can be found either by using components or by using a vector triangle. The vector triangle of velocities is shown in Figure 13.17.

Using Components

We use the formula

$$V_{BA} = V_B - V_A$$

$$= \begin{pmatrix} 0 \\ 40 \end{pmatrix} - \begin{pmatrix} 25 \\ 0 \end{pmatrix}$$

$$= \begin{pmatrix} -25 \\ 40 \end{pmatrix}$$

The magnitude of the relative velocity is given by

$$|V_{BA}| = \sqrt{(-25)^2 + 40^2}$$

$$= \sqrt{2225}$$

$$= 47.2$$

The velocity of B relative to A is 47.2 m s^{-1}.

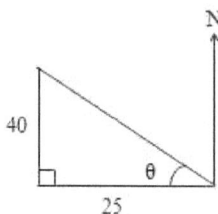

$$\tan\theta = \frac{40}{25}$$

$$\theta = 58°$$

The direction of the velocity is 328°.

Using Trigonometry

From the vector triangle, we have

$$V^2{}_{BA} = V^2{}_A + V^2{}_B$$

$$= 25^2 + 40^2$$

$$= 2225$$

$$|V_{AB}| = 47.2 \text{ m s}^{-1}$$

$$\tan\theta = \frac{25}{40}$$

$$\theta = 32°$$

The direction of the velocity is 328°

Exercise 13.3(a)

1. A moves at $3i + 5j$, and B moves at $2i + 3j$. Find the magnitude of the relative velocity of A to B, and the angle it makes with the x-axis.

2. A moves at $4i - 5j$, and the relative velocity of A to B is $2i + 3j$. Find the velocity of B in vector form. Find its magnitude and direction.

3. Ship A sails at 15 km h^{-1} on a bearing of 060°, and ship B sails at 12 km h^{-1} on a bearing of 330°. Find the relative velocity of A to B.

4. A fighter plane flies at 800 km h^{-1} on a bearing of 150°, and a bomber flies at 500 km h^{-1} on a bearing of 240°. Find the velocity of the fighter relative to the bomber.

5. A cruiser sails 200 km h^{-1} on a bearing of 080°, and a destroyer sails at 300 km h^{-1} on a bearing of 130°. Find the velocity of the cruiser relative to the destroyer.

6. A ship sailing due north at 15 km h^{-1}. The wind appears to come from 060° with speed 40 km h^{-1}. Find the true direction and speed of the wind.

Interception

Consider two particles A and B initially at points P and Q moving with constant velocity V_A and V_B respevtively. The particle A would intercept B if the relative velocity of A to B is along the straight line joining them.

Examples

1. A cruiser 60 km due north of a destroyer is travelling east at 18 km h^{-1}. The destroyer can travel at 22 km h^{-1}. Find the direction the destroyer sholud travel in order to intercept the cruiser. Find also to the nearest hour when the interception takes place.

    Let C and D stand for cruiser and destroyer. Figure 13.18 shows the relative position of the destroyer and cruiser. A vector of velocities is shown in Figure 13. 19.

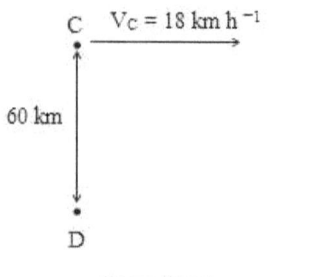

Figure 13.18   Figure 13.19

The destroyer should travel so that its velocity relative to the cruiser is due north. From the vector diagram, we have

$$\sin \theta = \frac{18}{22}$$

$$\theta = 54.9°$$

The destroyer should travel on a bearing of 054.9°.

$$V^2{}_{DC} = V^2{}_D - V^2{}_C$$

$$= 22^2 - 18^2$$

$$= 160$$

$$|V_{DC}| = 12.65$$

We use the average speed formula $s = d/t$.

$$t = \frac{60}{12.65}$$

$$= 4.74$$

Interception takes place after 5 hours.

2. Two ships A and B are 80 km apart with A due west of B. A is travelling at 16 km h^{-1} on a bearing of 040°. The ship B can travel at 25 km h^{-1}. Find the course that B should set in order to intercept A and the time when the interception takes place.

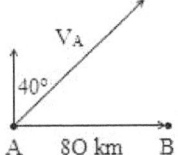
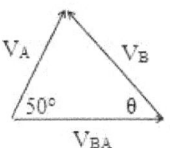

$$\frac{\sin\theta}{16} = \frac{\sin 50°}{25}$$

$$\sin\theta = \frac{16\sin 50°}{25}$$

$$\theta = 29.4°$$

The course set is 299.4°

$$\frac{V_{BA}}{\sin 100.6°} = \frac{25}{\sin 50°}$$

$$V_{BA} = \frac{25\sin 100.6°}{\sin 50°}$$

$$= 32.08$$

$$\text{Time} = \frac{50}{32.08}$$

$$= 1.56$$

Interception takes place after 1.56 hours.

Exercise 13.3(b)

1. A bomber is flying due west at 750 km h^{-1}. A fighter is 50 km due north of the bomber, and it sets off in pursuit at 1000 km h^{-1}. What course should the fighter pilot set, and how long will interception take?

2. A ship sails due north at 10 km h^{-1}. A submarine is 15 km away on a bearing of 030°. If the submarine can travel at 16 km h^{-1}, what course should be set to intercept the ship? How long will it be before interception?

3. Ship A is 80 km west of ship B. A sets off at 12 km h^{-1} on a course of 052°. If ship B can travel at 25 km h^{-1} find the course it should take to intercept A.

4. A bomber is flying due west at 50 km h^{-1}. A fighter 40 km due south of the bomber set off in pursuit at 800 km h^{-1}. What course should the fighter pilot set, and how long will interception take?

5. At 11:00 am two ships A and B are 15 km apart with B due east of A. A is travelling at 30 km h^{-1} in N60°E and B is travelling at 12 km h^{-1} due north. At what angle should B travel to intercept A and at what time would the interception occur?

6. Object B on the bearing of N10°E is 20 km from A. Object B is moving at a speed of 40 km h^{-1} on the bearing of 040°. If object A is capable of moving at 120 km h^{-1}, at what angle does it need to move to intercept B?

Review exercise 13

1. Given that $a = 3i + 2j$, $b = -i + 3j$ and $c = 2i - 3j$, find:

   (a) $a + 2b$ (b) $2c - 3b$ (c) $3a + 2c$

2. Given that $p = 3i + 2j$ and $q = 2i - 3j$, find the numbers x and y such that $xp + yq = 3i - 11j$.

3. If $a = i - 3j$, $b = 2i + 3j$ and $c = 2i + 4j$, find the value of the constants m and n such that $2a = 5nb - 3nc$.

4. Given that $a = -12i + 4j$ and $b = i + pj$, find the values of the constants p and q such that $a + qb = -27i + 19j$.

5. Given that $a = \begin{pmatrix} 3 \\ 1 \end{pmatrix}$, $b = \begin{pmatrix} -4 \\ 3 \end{pmatrix}$ and $c = \begin{pmatrix} 17 \\ -3 \end{pmatrix}$, find the value of m and of n such that $ma + nb = c$.

6. Given that $a = -2i + j$, $b = 3i + 2j$ and $c = -i - 2j$, find:

   (a) $|a + 2b|$ (b) $|2c - 3b|$ (c) $|3a - 2c|$

7. Find the magnitude of the following vectors.

   (a) $\begin{pmatrix} -3 \\ 4 \end{pmatrix}$ (b) $\begin{pmatrix} 5 \\ -12 \end{pmatrix}$ (c) $-2i + 3j$ (d) $4i - 5j$

8. Find the unit vectors in the direction of the following vectors.

   (a) $a = -3i + 4j$ (b) $b = 3i - j$ (c) $a = 2i + j$ (d) $b = -5i + 12j$

9. A(− 3, 2), B(1, − 2) and C(2, 3) are points in the x-y plane. Find $\vec{BA}$ and $\vec{BC}$ in the form $\begin{pmatrix} x \\ y \end{pmatrix}$.

10. A, B and C are points with position vectors $3i - 2j$, $2i + j$ and $-i + 4j$ respectively. Find in terms of i and j, the vectors $\vec{AB}$, $\vec{BC}$ and $\vec{CA}$.

11. The coordinates of A are (2, − 3) and the position vector of B is $4i + 2j$. Find the vector $\vec{BA}$.

12. If the coordinates of A are (4, 2) and $\vec{AB} = 2i + j$, find the position vector of B.

13. A and B have position vectors $2i + 5j$ and $4i - j$ respectively. Find the position vector of C if $3\vec{OA} = 2\vec{OB} + \vec{OC}$.

14. The coordinates of A and B are (3, 2) and (5, − 2) respectively. Find the position vector of C if $2\vec{OA} = 2\vec{OB} + \vec{BC}$.

15. The position vectors relative to an origin O, of three points A, B and C are $3i + 2j$, $7i + 4j$ and $2i + j$ respectively. Given that $\vec{OB} = m\vec{OA} + n\vec{OC}$, where m and n are scalar constants, find the value of m and of n.

16. The position vector of A and B are $2i - 3j$ and $-i - j$. Given that $\vec{AC} = i + 5j$, find the position vector of C. Find $|\vec{BC}|$ and the angle $\vec{BC}$ makes with the x-axis.

17. A quadrilateral has vertices A(4, 0), B(7, − 3), C(− 2, − 2) and D(− 5, 1). Show that ABCD is a parallelogram.

18. Show that the triangle whose vertices have position vectors $4i + 2j$, $2i + 5j$ and $5i + 3j$ is isosceles.

19. A(1, 2), B(3, 5), C(3, − 6) and D(x, y) are the vertices of the parallelogram ABCD. Find the value of x and of y.

20. P and Q divides the sides BC and AC respectively of triangle ABC in the ratio 2 : 1. If $\vec{AB} = a$ and $\vec{AC} = b$, (a) find $\vec{QP}$ and (b) show that $\vec{QP}$ is parallel to $\vec{AB}$ and one-third its length.

21. A pilot steered an aircraft on a bearing 060°. There was a 45 km h^{-1} wind blowing from the south-east. If the airspeed of the aircraft was 235 km h^{-1}, find the aircraft's track and its ground speed.

22. An airplane whose speed in still air is 498 km h^{-1} travels directly from A to B, distance of 1,600 km. The bearing of B from A is 330°, and there is a wind of 80 km h^{-1} from east.

   (a) Find the bearing on which the plane was steered.

   (b) Find time taken for the journey.

23. A motor boat can travel at 8 m s^{-1} in still water. The boat travels from P on one bank of a river flowing north at 5 m s^{-1}.

   (a) Find the speed of the boat in the water.

   (b) Find the angle to the bank that the boat should travel in order to reach a point Q directly opposite P.

24. A canoe travelling 12 m s^{-1} east encounters a current travelling 5 m s^{-1} north. The river is 90 metres wide.

   (a) What is the resultant velocity of the canoe?

   (b) How long does it take the canoe to travel to a point which is directly opposite its starting point?

25. A moves at $4i + 3j$, B moves at $3i + 7j$. Find the magnitude of the relative velocity of A to B, and the angle it makes with the x-axis.

26. Ship A sails at 13 km h^{-1} on a bearing of 050°, and ship B sails at 15 km h^{-1} on a bearing of 315°. Find the relative velocity of A to B.

27. A cruiser 80 km due north of a destroyer is travelling east at 15 km h^{-1}. The destroyer can travel at 25 km h^{-1}. Find the direction the destroyer should travel in order to intercept the cruiser. Find also to the nearest hour when the interception takes place.

28. Two ships A and B are 100 km apart with A due west of B. A is travelling at 18 km h^{-1} on a bearing of 050°. The ship B can travel at 32 km h^{-1}. Find the course that B should set in order to intercept A and the time when the interception takes place.

14 Matrices

14.1 Matrix Algebra

In a survey, 30 students said they like playing football of which 12 are girls, 25 like listing to music of which 15 are girls and 18 like playing computer games of which 6 are girls.

This information may be arranged in the rectangular form as shown below.

	Football	Music	Computer games
Number of girls	12	15	6
Number of boys	18	10	12

The same information may be written in brief as follows.

$$\begin{pmatrix} 12 & 15 & 6 \\ 18 & 10 & 12 \end{pmatrix}$$

A rectangular array of numbers arranged in rows and columns, and usually enclosed in brackets such as shown above is called a matrix. This matrix has 2 rows and 3 columns. The first row represents the number of girls and the second row represents the number of boys. The first, second and third columns represent the number of students who like football, music and computer games respectively. The plural of matrix is matrices.

The numbers of a matrix are called elements (or entries). A matrix that has m rows and n columns is called $m \times n$, read "m by n", matrix and is said to be of order (or size) $m \times n$.

Here are examples of matrices.

$$(5 \quad 6), \quad \begin{pmatrix} 4 \\ -3 \end{pmatrix}, \quad \begin{pmatrix} 2 & -3 \\ -1 & 0 \\ 3 & 4 \end{pmatrix} \quad \text{and} \quad \begin{pmatrix} 4 & 5 & 6 \\ -1 & 0 & -2 \\ 2 & -3 & 1 \end{pmatrix}$$

The first example has 1 row and 2 columns which mean that it is 1×2 matrix. The second example is 2×1 matrix. The third example is 3×2 matrix and the last example is 3×3 matrix

A matrix with 1 row of elements is called row matrix (or row vector), and a matrix with 1 column of elements is called column matrix (or column vector).

Square Matrix

A matrix that has the same number of rows as it has columns is called a square matrix. For example,

$$\begin{pmatrix} -3 & 1 \\ 2 & 0 \end{pmatrix}$$

is a 2 × 2 square matrix, and

$$\begin{pmatrix} 0 & 4 & -1 \\ -2 & 0 & 1 \\ 3 & 2 & 1 \end{pmatrix}$$

is a 3 × 3 square matrix.

Equal Matrices

Two matrices A and B are equal, written A = B, if they are of the same order and if corresponding elements are equal.

Example

Given that

$$\begin{pmatrix} 3x+y & 5 \\ -2 & -2x+y \end{pmatrix} = \begin{pmatrix} -1 & 5 \\ x & 3z \end{pmatrix},$$

find the values of x, y and z.

Because the two matrices are equal, we have

$$3x + y = -1$$

$$-2 = x$$

Substituting $x = -2$ into the first equation gives

$$3(-2) + y = -1$$

$$y = 5$$

Also, we have

$$-2x + y = 3z$$

Substituting -2 for x and 5 for y, we have

$$-2(-2) + 5 = 3z$$

giving
$$z = 3$$

Addition of Matrices

If A and B are two matrices of the same order, their sum A + B is obtained by adding corresponding elements of A and B.

Example

Given that $A = \begin{pmatrix} -1 & 2 \\ 9 & 4 \end{pmatrix}$ and $B = \begin{pmatrix} 2 & 3 \\ 4 & 8 \end{pmatrix}$, find A + B and B + A.

$$A + B = \begin{pmatrix} -1 & 2 \\ 9 & 4 \end{pmatrix} + \begin{pmatrix} 2 & 3 \\ 4 & 8 \end{pmatrix}$$

$$= \begin{pmatrix} 1 & 5 \\ 13 & 12 \end{pmatrix}$$

$$B + A = \begin{pmatrix} 2 & 3 \\ 4 & 8 \end{pmatrix} + \begin{pmatrix} -1 & 2 \\ 9 & 4 \end{pmatrix}$$

$$= \begin{pmatrix} 1 & 5 \\ 13 & 12 \end{pmatrix}$$

Observe that matrix addition is commutative.

Negative Matrices

If A is any matrix, the negative matrix of A, denoted by $-A$ is the matrix obtained by replacing each element in A by its negative.

Example

Given that $A = \begin{pmatrix} 3 & -2 \\ 0 & 4 \end{pmatrix}$, find $-A$.

$$-A = -\begin{pmatrix} 3 & -2 \\ 0 & 4 \end{pmatrix}$$

$$= \begin{pmatrix} -3 & 2 \\ 0 & -4 \end{pmatrix}$$

Subtraction of Matrices

If A and B are two matrices of the same order, the difference A − B is obtained by subtracting the elements in B from the corresponding elements in A.

Example

Given that $A = \begin{pmatrix} -2 & 3 \\ -8 & 7 \end{pmatrix}$ and $B = \begin{pmatrix} -2 & 4 \\ -10 & 6 \end{pmatrix}$,

find A − B and B − A.

$$A - B = \begin{pmatrix} -2 & 3 \\ -8 & 7 \end{pmatrix} - \begin{pmatrix} -2 & 4 \\ -10 & 6 \end{pmatrix}$$

$$= \begin{pmatrix} 0 & -1 \\ 2 & 1 \end{pmatrix}$$

$$B - A = \begin{pmatrix} -2 & 4 \\ -10 & 6 \end{pmatrix} - \begin{pmatrix} -2 & 3 \\ -8 & 7 \end{pmatrix}$$

$$= \begin{pmatrix} 0 & 1 \\ -2 & -1 \end{pmatrix}$$

Notice that A − B ≠ B − A.

Zero Matrix

A matrix in which all elements are zeros is called a zero matrix. The zero matrix is denoted by 0. For any matrix A, and the zero matrix 0 of the same order

$$A + 0 = 0 + A = A \text{ and } A - A = 0$$

Multiplying a Matrix by a Scalar

When a matrix is multiplied by a scalar k, each element of the matrix A is multiplied by k.

Example

Given that $A = \begin{pmatrix} -1 & 0 \\ 2 & -3 \end{pmatrix}$, find −2A.

$$-2A = -2 \begin{pmatrix} -1 & 0 \\ 2 & -3 \end{pmatrix} = \begin{pmatrix} 2 & 0 \\ -4 & 6 \end{pmatrix}$$

Exercise 14.1(a)

1. Find the values of x, y and z.

   (a) $\begin{pmatrix} 2 & x+y \\ 2y & 6 \end{pmatrix} = \begin{pmatrix} x-y & 4 \\ z & 6 \end{pmatrix}$

   (b) $\begin{pmatrix} 3 & x-2y \\ 5 & -2 \end{pmatrix} = \begin{pmatrix} 3 & -4 \\ x+y & -2 \end{pmatrix}$

2. Find:

   (a) $\begin{pmatrix} 5 & 3 \\ 1 & 2 \end{pmatrix} + \begin{pmatrix} 6 & 4 \\ 1 & 3 \end{pmatrix}$

   (b) $\begin{pmatrix} 0 & 2 \\ 4 & -1 \end{pmatrix} + \begin{pmatrix} 6 & -1 \\ 5 & 2 \end{pmatrix}$

   (c) $\begin{pmatrix} 4 & 2 \\ 3 & -1 \end{pmatrix} + \begin{pmatrix} 1 & 3 \\ -2 & 2 \end{pmatrix}$

   (d) $\begin{pmatrix} 5 & 3 \\ 2 & 4 \end{pmatrix} + \begin{pmatrix} -2 & 2 \\ -1 & 0 \end{pmatrix}$

   (e) $\begin{pmatrix} 3 & -1 & 0 \\ -2 & 0 & 1 \end{pmatrix} + \begin{pmatrix} -5 & 2 & -1 \\ 4 & 3 & -4 \end{pmatrix}$

   (f) $\begin{pmatrix} -4 & 2 & 3 \\ 5 & -3 & 1 \end{pmatrix} + \begin{pmatrix} 5 & -3 & -2 \\ -3 & 7 & -3 \end{pmatrix}$

   (g) $\begin{pmatrix} 0 & 5 \\ 3 & -4 \\ -2 & 1 \end{pmatrix} + \begin{pmatrix} 1 & -6 \\ 2 & 6 \\ 3 & -2 \end{pmatrix}$

   (h) $\begin{pmatrix} -7 & 8 & 6 \\ 5 & 4 & -3 \\ 2 & -5 & 1 \end{pmatrix} + \begin{pmatrix} 9 & -6 & -10 \\ -3 & 0 & 5 \\ -3 & 6 & -2 \end{pmatrix}$

3. Find the values of a, b, c and d so that

   $\begin{pmatrix} a & b \\ c & d \end{pmatrix} + \begin{pmatrix} -2 & 3 \\ 2 & 1 \end{pmatrix} = \begin{pmatrix} -1 & 2 \\ -3 & 4 \end{pmatrix}$

4. Find the values of w, x, y and z so that

   $\begin{pmatrix} -1 & 5 \\ 1 & -2 \end{pmatrix} + \begin{pmatrix} w & x \\ y & z \end{pmatrix} = \begin{pmatrix} -2 & 3 \\ 6 & 1 \end{pmatrix}$

5. Find the values of x and y so that

$$\begin{pmatrix} 2x & 5 \\ -1 & 3x \end{pmatrix} + \begin{pmatrix} 3y & -3 \\ -5 & -2y \end{pmatrix} = \begin{pmatrix} 7 & 2 \\ -6 & 4 \end{pmatrix}$$

6. Find the values of x and y so that

$$\begin{pmatrix} -3x & -3 \\ 2 & 4x \end{pmatrix} + \begin{pmatrix} 2y & 5 \\ -2 & -5y \end{pmatrix} = \begin{pmatrix} -1 & 2 \\ 0 & -1 \end{pmatrix}$$

7. Find the negate matrix of each of the following matrices.

(a) $\begin{pmatrix} -2 & 3 \\ 1 & -4 \end{pmatrix}$

(b) $\begin{pmatrix} -3 & -2 \\ 5 & -1 \end{pmatrix}$

(c) $\begin{pmatrix} 6 & 0 \\ 0 & -6 \end{pmatrix}$

(d) $\begin{pmatrix} 3 & -2 & 0 \\ -1 & 5 & -2 \end{pmatrix}$

(e) $\begin{pmatrix} -4 & 0 \\ 3 & 1 \\ -2 & -1 \end{pmatrix}$

(f) $\begin{pmatrix} 2 & -5 & 3 \\ -3 & 0 & -2 \\ 4 & -3 & 2 \end{pmatrix}$

8. Find:

(a) $\begin{pmatrix} -1 & 1 \\ 4 & 2 \end{pmatrix} - \begin{pmatrix} -2 & 0 \\ 4 & 1 \end{pmatrix}$

(b) $\begin{pmatrix} 4 & 2 \\ -1 & 3 \end{pmatrix} - \begin{pmatrix} 1 & 8 \\ 1 & 0 \end{pmatrix}$

(c) $\begin{pmatrix} 3 & -6 \\ -2 & 1 \end{pmatrix} - \begin{pmatrix} -4 & -2 \\ 1 & -3 \end{pmatrix}$

(d) $\begin{pmatrix} -2 & 4 \\ -8 & 5 \end{pmatrix} - \begin{pmatrix} -2 & 0 \\ -5 & 3 \end{pmatrix}$

(e) $\begin{pmatrix} 0 & 3 & -1 \\ 1 & -2 & 0 \end{pmatrix} - \begin{pmatrix} -1 & -5 & 2 \\ -4 & 4 & 3 \end{pmatrix}$

(f) $\begin{pmatrix} 5 & -3 & 1 \\ -4 & 2 & 1 \end{pmatrix} - \begin{pmatrix} -3 & -7 & -3 \\ -5 & 3 & 0 \end{pmatrix}$

(g) $\begin{pmatrix} 5 & 0 \\ -4 & 3 \\ 1 & -2 \end{pmatrix} - \begin{pmatrix} -6 & -1 \\ 6 & 2 \\ -2 & 3 \end{pmatrix}$

(h) $\begin{pmatrix} 2 & -5 & 1 \\ -7 & 8 & 6 \\ 2 & -5 & 1 \end{pmatrix} - \begin{pmatrix} -3 & -6 & 2 \\ -3 & 6 & 4 \\ -5 & 7 & -2 \end{pmatrix}$

9. Find:

(a) $3 \begin{pmatrix} 1 & 2 \\ -2 & 1 \end{pmatrix}$

(b) $-2 \begin{pmatrix} -3 & 1 \\ 1 & -2 \end{pmatrix}$

(c) $4 \begin{pmatrix} -1 & 0 \\ 0 & 1 \end{pmatrix}$

(d) $-\dfrac{1}{2}\begin{pmatrix} -2 & 6 \\ 4 & 0 \end{pmatrix}$ (e) $\dfrac{2}{3}\begin{pmatrix} -6 & 0 \\ 12 & -9 \end{pmatrix}$

(f) $-\dfrac{3}{2}\begin{pmatrix} 4 & -2 \\ -6 & 2 \end{pmatrix}$ (g) $-2\begin{pmatrix} 2 & -1 & -3 \\ 0 & 3 & -2 \end{pmatrix}$

(h) $3\begin{pmatrix} 0 & 1 \\ -1 & 2 \\ -2 & 3 \end{pmatrix}$ (i) $-\dfrac{1}{3}\begin{pmatrix} 3 & 0 & -6 \\ -3 & 9 & -3 \\ 0 & -3 & 6 \end{pmatrix}$

10. Given that

$$A = \begin{pmatrix} -3 & 2 \\ 2 & 0 \end{pmatrix}, B = \begin{pmatrix} -2 & 1 \\ 1 & 5 \end{pmatrix} \text{ and } C = \begin{pmatrix} 0 & -3 \\ 1 & 2 \end{pmatrix},$$

find :

(a) $2A + B$ (b) $3C - 2B$

(c) $2C - 3A$ (d) $2(B - A) + C$

(e) $3B - 2A + B$

11. A shop sells two brands, A and B, of refrigerators, stoves and microwave ovens. One week the shop sold 25 brand A and 30 brand B refrigerators, 18 brand A and 45 brand B stoves and 12 brand A and 28 brand B microwave ovens. Write a 2 × 3 matrix to display this information.

12. In a survey of 2000 students who apply for university admission, 900 were admitted to the Art faculty, of which 65% were female, 700 were admitted to the Science faculty, of which 40% were female and the remaining were admitted to the Education faculty, of which 54% were female. Write a 3 × 2 matrix to display this information.

13. During March shop A sold 120 shirts and 80 pairs of trousers, while shop B sold 150 shirts and 65 pairs of trousers. Shop A sold 165 shirts and 95 pairs of trousers in April, while shop B sold 126 shirts and 78 pairs of trousers. Use matrix addition to find the total number of shirts and pair of trousers sold over the 2-month period for each shop.

14. In 2018, a university admitted 120 males and 80 females to the law faculty and 80 males and 70 females to the science faculty. In 2019, the university admitted 150 males and 95 females to the law faculty and 98 males and 86 females to the science faculty. Use matrix subtraction to find the increase in the males and females admitted to each faculty.

Multiplication of Matrices

Multiplication of two matrices is possible only if the number of columns of the first matrix equals the number of rows of the second matrix. For example, a 2 × 3 matrix can be multiplied with a 3 × 2 matrix but a 2 × 3 matrix cannot be multiplied with 4 × 3 matrix.

The number of rows of the product of two matrices equals the number of rows of the first matrix, and the number of columns of the second matrix. For example, the product of 2 × 3 matrix with 3 × 4 matrix would be a 2 × 4 matrix.

If A and B are two 2 × 2 matrices, the product AB is obtained as follows:

Multiply 1st element in the 1st row in A by 1st element in the 1st column in B. Next, multiply 2nd element in the 1st row in A by the 2nd element in 1st column in B. Then add the two results to give the 1st element in the 1st row and 1st column of the matrix AB. To obtain the element in the 1st row and 2nd column of AB, first multiply the 1st element in the 1st row in A by the 1st element in the second column in B. Next, multiply the 2nd element in the 1st row by the 2nd element in 2nd column. Finally add the two results. The second row of AB is obtained in the same manner, using the second row of matrix A. The following example illustrates the steps.

Example

Given that

$$A = \begin{pmatrix} 2 & 0 \\ -3 & 1 \end{pmatrix} \text{ and } B = \begin{pmatrix} -1 & 3 \\ -2 & 4 \end{pmatrix},$$

find AB and BA.

$$AB = \begin{pmatrix} 2 & 0 \\ -3 & 1 \end{pmatrix} \begin{pmatrix} -1 & 3 \\ -2 & 4 \end{pmatrix}$$

$$= \begin{pmatrix} 2(-1) + (0)(-2) & 2(3) + (0)(4) \\ -3(-1) + (1)(-2) & -3(3) + (1)(4) \end{pmatrix}$$

$$= \begin{pmatrix} -2 & 6 \\ 1 & -5 \end{pmatrix}$$

$$BA = \begin{pmatrix} -1 & 3 \\ -2 & 4 \end{pmatrix} \begin{pmatrix} 2 & 0 \\ -3 & 1 \end{pmatrix}$$

$$= \begin{pmatrix} (-1)(2) + (3)(-3) & (-1)(0) + (3)(1) \\ (-2)(2) + (4)(-3) & (-2)(0) + (4)(1) \end{pmatrix}$$

$$= \begin{pmatrix} -11 & 3 \\ -16 & 4 \end{pmatrix}$$

Observe that AB ≠ BA. Matrix multiplication is not commutative.

If A is a square matrix then AA is defined and is denoted by A^2. Similarly, $AAA = A^3$.

It is possible to find two matrices A and B, with non-zero elements such that AB = 0. For example,

$$\begin{pmatrix} -2 & 3 \\ -4 & 6 \end{pmatrix} \begin{pmatrix} 3 & -6 \\ 2 & -4 \end{pmatrix} = \begin{pmatrix} 0 & 0 \\ 0 & 0 \end{pmatrix}$$

Exercise 14.1(b)

1. Find:

(a) $\begin{pmatrix} 1 & 2 \\ 0 & 3 \end{pmatrix} \begin{pmatrix} 3 & 2 \\ 4 & -1 \end{pmatrix}$ (b) $\begin{pmatrix} 2 & -3 \\ -5 & 1 \end{pmatrix} \begin{pmatrix} 1 & 2 \\ 3 & 2 \end{pmatrix}$

(c) $\begin{pmatrix} 1 & 3 \\ -2 & 5 \end{pmatrix} \begin{pmatrix} 0 & -2 \\ 2 & -4 \end{pmatrix}$ (d) $\begin{pmatrix} 3 & 4 \\ 0 & 1 \end{pmatrix} \begin{pmatrix} 6 & 2 \\ 5 & 0 \end{pmatrix}$

(e) $\begin{pmatrix} 6 & -3 \\ -4 & 2 \end{pmatrix} \begin{pmatrix} 4 & 3 \\ 8 & 0 \end{pmatrix}$ (f) $\begin{pmatrix} 3 & 2 \\ -3 & -2 \end{pmatrix} \begin{pmatrix} 2 & -4 \\ -3 & 6 \end{pmatrix}$

(g) $\begin{pmatrix} 1 & 2 \\ 2 & 4 \end{pmatrix} \begin{pmatrix} 2 \\ 3 \end{pmatrix}$ (h) $\begin{pmatrix} 1 & 3 \\ 2 & 5 \end{pmatrix} \begin{pmatrix} 2 \\ -1 \end{pmatrix}$

(i) $\begin{pmatrix} 1 & -3 \\ -2 & 1 \end{pmatrix} \begin{pmatrix} 1 \\ 5 \end{pmatrix}$ (j) $\begin{pmatrix} 4 & -3 \\ -2 & 1 \end{pmatrix} \begin{pmatrix} -4 \\ -5 \end{pmatrix}$

(k) $\begin{pmatrix} 3 & -2 & 1 \\ 1 & 0 & 2 \end{pmatrix} \begin{pmatrix} -1 & 0 \\ 2 & 1 \\ -3 & -2 \end{pmatrix}$ (l) $\begin{pmatrix} 4 & -2 \\ 2 & 1 \\ -1 & 0 \end{pmatrix} \begin{pmatrix} 3 & -1 \\ 2 & 0 \end{pmatrix}$

(m) $\begin{pmatrix} -2 & 1 & -1 \\ 3 & 4 & 0 \\ -1 & 2 & -2 \end{pmatrix} \begin{pmatrix} 0 & 2 \\ 1 & -2 \\ 0 & 1 \end{pmatrix}$ (n) $\begin{pmatrix} 4 & 5 & -3 \\ 0 & 2 & 1 \\ -1 & 0 & 2 \end{pmatrix} \begin{pmatrix} 0 & 1 & 2 \\ 4 & -1 & 0 \\ 3 & 2 & 1 \end{pmatrix}$

CHAPTER 14 Matrices

2. Find A^2 and A^3 for each of the following matrices.

   (a) $\begin{pmatrix} 1 & -1 \\ 3 & 2 \end{pmatrix}$ (b) $\begin{pmatrix} -3 & -1 \\ 2 & 1 \end{pmatrix}$ (c) $\begin{pmatrix} 1 & 0 \\ 2 & -1 \end{pmatrix}$

3. Find the values of x, y and z so that

$$\begin{pmatrix} 2 & 3 \\ -3 & 1 \end{pmatrix} \begin{pmatrix} x & 2 \\ -1 & y \end{pmatrix} = \begin{pmatrix} 1 & z \\ -7 & z \end{pmatrix}$$

4. Find the values of x, y and z so that

$$\begin{pmatrix} y & 3 \\ x & 1 \end{pmatrix} \begin{pmatrix} 2 & 3 \\ 1 & -2 \end{pmatrix} = \begin{pmatrix} -1 & -12 \\ z & z \end{pmatrix}$$

5. Find the values of a, b, c and d so that

$$\begin{pmatrix} 2 & 3 \\ 3 & -2 \end{pmatrix} \begin{pmatrix} a & b \\ c & d \end{pmatrix} = \begin{pmatrix} -7 & 8 \\ -4 & -1 \end{pmatrix}$$

6. Find the values of a, b, c and d so that

$$\begin{pmatrix} -3 & 2 \\ 1 & -2 \end{pmatrix} \begin{pmatrix} a & b \\ c & d \end{pmatrix} = \begin{pmatrix} 0 & 5 \\ -4 & 1 \end{pmatrix}$$

7. Shop A sold 180 shirts and 90 pairs of trousers in a month, while shop B sold 165 shirts and 126 pair of trousers. The shops make £6 profit on each shirt and £8 profit on each pair of trousers. Use matrix multiplication to find the profit made in the month by each shop.

8. John went to a store and purchased 8 textbooks and 12 exercise books. Eric purchased 6 textbooks and 15 exercise books. If the textbooks cost £16 each and the exercise books cost £3.50 each, use matrix multiplication to find the amount spent by John and Eric.

Identity Matrix

A square matrix whose leading diagonal consists of 1s where all the rest of the matrix consists of zeros is called the identity (or unit) matrix. For example,

$$\begin{pmatrix} 1 & 0 \\ 0 & 1 \end{pmatrix}$$

is the 2 × 2 identity matrix and

$$\begin{pmatrix} 1 & 0 & 0 \\ 0 & 1 & 0 \\ 0 & 0 & 1 \end{pmatrix}$$

is the 3 × 3 identity matrix.

The identity matrix is denoted by I.

If A is a square matrix of order n and I the identity matrix of the same order then IA = AI = A.

Example

Given that $A = \begin{pmatrix} -2 & 3 \\ 1 & 4 \end{pmatrix}$ find AI and IA.

$$AI = \begin{pmatrix} -2 & 3 \\ 1 & 4 \end{pmatrix} \begin{pmatrix} 1 & 0 \\ 0 & 1 \end{pmatrix}$$

$$= \begin{pmatrix} -2 & 3 \\ 1 & 4 \end{pmatrix}$$

$$IA = \begin{pmatrix} 1 & 0 \\ 0 & 1 \end{pmatrix} \begin{pmatrix} -2 & 3 \\ 1 & 4 \end{pmatrix}$$

$$= \begin{pmatrix} -2 & 3 \\ 1 & 4 \end{pmatrix}$$

Determinants

Associated with each square matrix is a real number, called the determinant. A determinant of a matrix A, denoted by |A| or det A, is defined for a 2 × 2 matrix as follows.

If $A = \begin{pmatrix} a & b \\ c & d \end{pmatrix}$ then $|A| = ad - bc$.

Example

Find the determinant of the matrix $\begin{pmatrix} 3 & -4 \\ -2 & 2 \end{pmatrix}$.

$$\begin{vmatrix} 3 & -4 \\ -2 & 2 \end{vmatrix} = 3(2) - (-4)(-2)$$

$$= -2$$

The Inverse of a Matrix

If A and B are two square matrices of the same order, then A and B are inverse of each other if AB = BA = I, where I is the identity matrix.

The inverse of a matrix A is denoted by A^{-1}, and read "A inverse".

A square matrix may or may not have an inverse. A matrix which has an inverse is called a non-singular matrix.

If $A = \begin{pmatrix} a & b \\ c & d \end{pmatrix}$ and $|A| \neq 0$, then

$$A^{-1} = \frac{1}{|A|}\begin{pmatrix} d & -b \\ -c & a \end{pmatrix}.$$

Example

Find the inverse of the matrix $A = \begin{pmatrix} 3 & 2 \\ 1 & 4 \end{pmatrix}$.

Begin by finding the determinant of A.

$$\begin{vmatrix} 3 & 2 \\ 1 & 4 \end{vmatrix} = 3(4) - 2(1)$$

$$= 10$$

$$A^{-1} = \frac{1}{10}\begin{pmatrix} 4 & -2 \\ -1 & 3 \end{pmatrix}$$

$$= \begin{pmatrix} 0.4 & -0.2 \\ -0.1 & 0.3 \end{pmatrix}$$

Exercise 14.1(c)

1. Evaluate:

   (a) $\begin{vmatrix} 4 & 3 \\ 3 & 2 \end{vmatrix}$ (b) $\begin{vmatrix} 2 & 1 \\ 4 & 3 \end{vmatrix}$

   (c) $\begin{vmatrix} -2 & 3 \\ -2 & 2 \end{vmatrix}$ (d) $\begin{vmatrix} -3 & -2 \\ 4 & 3 \end{vmatrix}$

   (e) $\begin{vmatrix} 6 & 3 \\ 2 & 1 \end{vmatrix}$ (f) $\begin{vmatrix} 6 & 4 \\ 5 & 3 \end{vmatrix}$

2. If $A = \begin{pmatrix} -2 & 1 \\ 1 & 3 \end{pmatrix}$ and $B = \begin{pmatrix} -1 & 2 \\ 5 & 4 \end{pmatrix}$, find:

   (a) $|A|$ (b) $|B|$ (c) $|AB|$

   (d) $|BA|$ (e) $|A||B|$

3. Solve for x.

   (a) $\begin{vmatrix} x & x \\ 3 & 4 \end{vmatrix} = -5$

   (b) $\begin{vmatrix} 1 & x \\ x & 3 \end{vmatrix} = 2$

   (c) $\begin{vmatrix} 4 & 5-x \\ 2-x & 1 \end{vmatrix} = 0$

4. Find the inverse of each of the following matrices.

   (a) $\begin{pmatrix} -2 & 3 \\ 2 & 0 \end{pmatrix}$ (b) $\begin{pmatrix} 4 & 5 \\ 2 & 3 \end{pmatrix}$

   (c) $\begin{pmatrix} -2 & -1 \\ 4 & 3 \end{pmatrix}$ (d) $\begin{pmatrix} 2 & 4 \\ 3 & 5 \end{pmatrix}$

   (e) $\begin{pmatrix} 5 & -2 \\ 0 & 2 \end{pmatrix}$ (f) $\begin{pmatrix} 3 & 2 \\ -1 & 2 \end{pmatrix}$

5. Given that $A = \begin{pmatrix} 1 & 2 \\ 2 & -1 \end{pmatrix}$ and $B = \begin{pmatrix} 1 & 4 \\ 3 & 1 \end{pmatrix}$ find:

   (a) A^{-1} (b) B^{-1}

   (c) $B^{-1}B$ (d) AA^{-1}

6. Given that $A = \begin{pmatrix} 5 & 4 \\ 3 & 2 \end{pmatrix}$ and $B = \begin{pmatrix} 8 & 4 \\ -3 & -2 \end{pmatrix}$ find:

   (a) A^{-1} (b) BA^{-1}

   (c) B^{-1} (d) $B^{-1}A$

14.2 Solving Simultaneous Equations

The inverse of a matrix can be used to solve simultaneous equations.

CHAPTER 14 Matrices

The simultaneous equations

$$a_{11}x + a_{12}y = b_1$$

$$a_{21}x + a_{22}y = b_2$$

can be written in the matrix form

$$AX = B$$

where $A = \begin{pmatrix} a_{11} & a_{12} \\ a_{21} & a_{22} \end{pmatrix}$, $X = \begin{pmatrix} x \\ y \end{pmatrix}$ and $B = \begin{pmatrix} b_1 \\ b_2 \end{pmatrix}$

If A^{-1} exists, then multiplying both sides of the equation by A^{-1}, we obtain

$$A^{-1}(AX) = A^{-1}B$$

$$(A^{-1}A)X = A^{-1}B$$

$$IX = A^{-1}B$$

$$X = A^{-1}B$$

It follows that, the simultaneous equations

$$AX = B$$

for which A is a square matrix and A^{-1} exists, has a unique solution given by

$$X = A^{-1}B$$

Example

Solve the simultaneous equations

$$x + 2y = 7$$

$$2x + 3y = 12$$

Here $A = \begin{pmatrix} 1 & 2 \\ 2 & 3 \end{pmatrix}$, $X = \begin{pmatrix} x \\ y \end{pmatrix}$ and $B = \begin{pmatrix} 7 \\ 12 \end{pmatrix}$.

Now $\begin{vmatrix} 1 & 2 \\ 2 & 3 \end{vmatrix} = 3 - 4 = -1$

Hence $A^{-1} = \begin{pmatrix} -3 & 2 \\ 2 & -1 \end{pmatrix}$

Solving Simultaneous Equations 221

Multiplying both sides of the matrix equation by A^{-1} gives

$$\begin{pmatrix} x \\ y \end{pmatrix} = \begin{pmatrix} -3 & 2 \\ 2 & -1 \end{pmatrix} \begin{pmatrix} 7 \\ 12 \end{pmatrix}$$

$$= \begin{pmatrix} 3 \\ 2 \end{pmatrix}$$

So, $x = 3$ and $y = 2$.

Exercise 14.2

1. Use matrix to solve the following simultaneous equations.

   (a) $5x + 4y = 3$

   $3x + 2y = 1$

   (b) $x + 2y = 1$

   $4x + y = 4$

   (c) $-x + 2y = 4$

   $-2x + 3y = 5$

   (d) $7x + 3y = -8$

   $5x + 2y = -6$

   (e) $3x + 2y = 5$

   $-2x + 3y = -12$

   (f) $2x - 3y = -5$

   $-x + 2y = 4$

2. Kojo has 22 coins with a total value of $1.70. If the coins are all $0.10 and $0.05, how many of each type of coin does he have?

3. 500 tickets were sold for a concert. The receipts from ticket sales were £3100 and the ticket prices were £5 and £8. How many of each price ticket were sold?

4. A man invests a part of £800 in bonds paying 12 percent interest. The remainder is in a savings account at 8 percent. If he receives £84 in interest for 1 year how much does he have invested at each rate?

5. A chemist has 15% and 25% acid solution. How much of each solution should be used to form 500 ml of a 21% acid solution.

6. A plane flies 540 km with the wind in 3 hours. Flying back against the wind, the plane takes 9 hours to make the trip. What was the rate of the plane in still air? What was the rate of the wind?

Review Exercise 14

1. Use the following matrices to compute the given expressions.

$$A = \begin{pmatrix} 1 & 2 \\ 0 & 4 \end{pmatrix} \quad B = \begin{pmatrix} 3 & -1 \\ 4 & 2 \end{pmatrix} \quad C = \begin{pmatrix} 2 & 3 \\ 4 & -2 \end{pmatrix}$$

(a) $A + B$ (b) $B + C$ (c) $2A - 3C$

(d) $A + C$ (e) $A - 2C$ (f) $3C - 4B$

(g) $(A + B) - 2C$ (h) $4C + (A - B)$ (i) $2A - 5(C + B)$

(j) $2(A - B) - C$ (k) $2(B - C) - A$ (l) $2(B - A) - C$

2. Use the following matrices to compute the given expressions.

$$A = \begin{pmatrix} 3 & 2 \\ -1 & 4 \end{pmatrix} \quad B = \begin{pmatrix} 2 & 4 \\ 3 & -1 \end{pmatrix} \quad C = \begin{pmatrix} 2 & 0 \\ 4 & -2 \end{pmatrix} \quad D = \begin{pmatrix} 2 \\ 3 \end{pmatrix}$$

(a) AB (b) CD (c) BC

(d) $AD - BD$ (e) $(A + I)C$

(f) $(B - C)D$ (g) $B(A + C)$

(h) $A(B - C)$ (i) $2BD - AD$

3. Given that the matrix $A = \begin{pmatrix} -3 & 1 \\ 0 & 2 \end{pmatrix}$, find:

(a) A^2 (b) A^3

4. Given that $A = \begin{pmatrix} 4 & -2 \\ 3 & -1 \end{pmatrix}$ and $B = \begin{pmatrix} 1 & 2 \\ -2 & 1 \end{pmatrix}$, find:

(a) A^2 (b) B^2

(c) $A^2 B$ (d) $B^2 A$

5. Find the values of a, b, c and d if: $\begin{pmatrix} 1 & 3 \\ 1 & 4 \end{pmatrix} \begin{pmatrix} a & b \\ c & d \end{pmatrix} = \begin{pmatrix} 6 & -5 \\ 7 & -7 \end{pmatrix}$

6. Find the values of x and y if: $\begin{pmatrix} -1 & -1 \\ 4 & 2 \end{pmatrix} \begin{pmatrix} x \\ y \end{pmatrix} + \begin{pmatrix} -1 \\ 2 \end{pmatrix} = \begin{pmatrix} -5 \\ 0 \end{pmatrix}$

7. If $A = \begin{pmatrix} 6 & 2 \\ 4 & y \end{pmatrix}$ and $B = \begin{pmatrix} 4 & 2 \\ x & 3 \end{pmatrix}$, find the value of x and of y, given that A and B are commutative under matrix multiplication.

8. Find the value of each determinant.

   (a) $\begin{vmatrix} 3 & 1 \\ 4 & 2 \end{vmatrix}$ (b) $\begin{vmatrix} 6 & 1 \\ 5 & 2 \end{vmatrix}$

   (c) $\begin{vmatrix} 8 & -3 \\ 4 & 2 \end{vmatrix}$ (d) $\begin{vmatrix} 3 & 2 \\ -1 & 4 \end{vmatrix}$

   (e) $\begin{vmatrix} 4 & -1 \\ 6 & -1 \end{vmatrix}$ (f) $\begin{vmatrix} 2 & 0 \\ 4 & -3 \end{vmatrix}$

9. Solve for x.

   (a) $\begin{vmatrix} x & x \\ 4 & 3 \end{vmatrix} = 5$ (b) $\begin{vmatrix} x & 1 \\ 3 & x \end{vmatrix} = -2$

   (c) $\begin{vmatrix} 5-x & 4 \\ 1 & 2-x \end{vmatrix} = 0$

   (d) $\begin{vmatrix} 2 & x \\ -1 & x^2 - 3x - 6 \end{vmatrix} = 0$

10. Find the inverse of each matrix.

    (a) $\begin{pmatrix} 2 & 1 \\ 1 & 1 \end{pmatrix}$ (b) $\begin{pmatrix} 3 & -1 \\ -2 & 1 \end{pmatrix}$ (c) $\begin{pmatrix} 6 & 5 \\ 2 & 2 \end{pmatrix}$

    (d) $\begin{pmatrix} -4 & 1 \\ 6 & -2 \end{pmatrix}$ (e) $\begin{pmatrix} -1 & -2 \\ 3 & 4 \end{pmatrix}$ (f) $\begin{pmatrix} 1 & 2 \\ 2 & 3 \end{pmatrix}$

11. Given that $A = \begin{pmatrix} -3 & 2 \\ -2 & 1 \end{pmatrix}$, find:

    (a) A + 3I   (b) 2I − A

12. Given that $A = \begin{pmatrix} 3 & 2 \\ 1 & 4 \end{pmatrix}$, find the matrix B such that $A^2B = I$.

13. Given that $A = \begin{pmatrix} 4 & 5 \\ -2 & -1 \end{pmatrix}$ and $B = \begin{pmatrix} -3 & 1 \\ 4 & -2 \end{pmatrix}$, find the matrix X such that

    (a) (A + B)X = I   (b) (A − B)X = I

CHAPTER 14 Matrices

14. Given a 2 × 2 matrix A such that $A^2 - 5A - 2I = 0$, show that $A^{-1} = \frac{1}{2}(A - 5I)$. Hence, find A^{-1} if $A = \begin{pmatrix} 1 & 2 \\ 3 & 4 \end{pmatrix}$.

15. Given that $A = \begin{pmatrix} 2 & 1 \\ -1 & 3 \end{pmatrix}$, evaluate A^{-1} and show that $A + 7A^{-1} = 5I$, where I is 2 × 2 identity matrix. Deduce that $A^2 - 5A + 7I = 0$ and use this equation to find A^2.

16. Solve each of the following simultaneous equations.

   (a) $5x + 2y = 7$

   $2x - y = 13$

   (b) $5x - y = 13$

   $2x + 3y = 12$

   (c) $x + 3y = 5$

   $2x - 3y = -8$

   (d) $3x + 7y = 13$

   $2x + 5y = 9$

17. Given that the matrix $A = \begin{pmatrix} -3 & 4 \\ 3 & -2 \end{pmatrix}$, find a 2 × 2 matrix X such that $AX = XA = I$, where I is the unit matrix. Hence, solve the simultaneous equations.

   (a) $-3x + 4y = -1$

   $3x - 2y = 5$

   (b) $-3x + 4y = 9$

   $3x - 2y = -7$

18. Given that $A = \begin{pmatrix} 4 & 3 \\ 3 & -2 \end{pmatrix}$, find the inverse matrix A^{-1} and hence solve the simultaneous equations.

   (a) $4x + 3y = 1$

   $3x - 2y = -12$

   (b) $4x + 3y = 11$

   $3x - 2y = 4$

15 Differentiation and Integration

15.1 Differentiation

Gradient of a Curve

The gradient of a straight line is constant but the gradient of a curve varies along its length. Therefore the gradient of a curve is not defined between two points as in the case of a straight line.

The gradient of a curve at any point is defined as the gradient of the tangent to the curve at that point. The tangent to a curve is a line that touches the curve at only one point, as illustrated in Figure 15.1. PQ is a tangent to the curve at P.

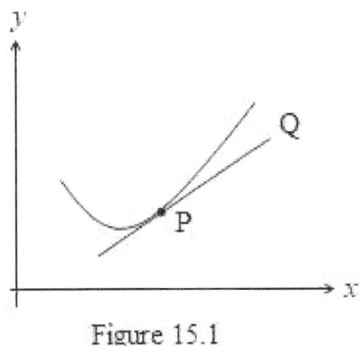

Figure 15.1

The gradient of the curve at P is equal to the gradient of line PQ.

The Derivative

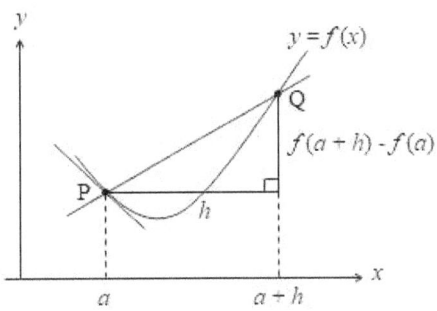

Figure 15.2

CHAPTER 15 Differentiation and Integration

Let P and Q be two points with coordinates $(a, f(a))$ and $(a + h, f(a + h))$ respectively on the graph of a function $y = f(x)$, as shown in Figure 15.2. Draw a line through P and Q, this line is called a secant. The gradient of the secant PQ is given by

$$\frac{\Delta y}{\Delta x} = \frac{f(a+h)-f(a)}{(a+h)-a}$$

$$= \frac{f(a+h)-f(a)}{h}$$

The gradient corresponds to the average rate of change of y with respect to x over the interval from a to $a + h$.

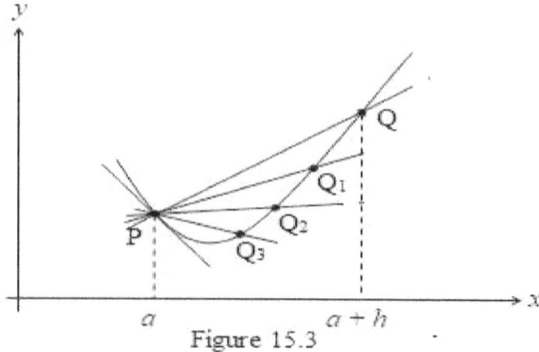

Figure 15.3

The point Q slides along the curve and gets closer and closer to the point P as h approaches 0, as shown in Figure 15.3. The points Q_1, Q_2 and Q_3 are successive position of Q. As h approaches 0, the point Q gets closer and closer to the point P. The corresponding gradients of the secant lines approach a limit as h approaches 0. This limit is defined to be the gradient of the tangent at P.

The tangent of the curve of the function $y = f(x)$ at the point $(a, f(a))$ is a line through this point having gradient

$$\lim_{h \to 0} \frac{f(a+h) - f(a)}{h}$$

The gradient of the tangent corresponds to the instantaneous rate of change of y with respect to x at the point. The gradient of the tangent at a point is also called the gradient of the curve at the point.

Consider the function $f(x) = x^2$. The gradient of the tangent at a point is obtained as follows:

Differentiation

$$\lim_{h \to 0} \frac{f(x+h) - f(x)}{h} = \lim_{h \to 0} \frac{(x+h)^2 - x^2}{h}$$

$$= \lim_{h \to 0} \frac{x^2 + 2xh + h^2 - x^2}{h}$$

$$= \lim_{h \to 0} \frac{2xh + h^2}{h}$$

$$= \lim_{h \to 0} (2x + h)$$

$$= 2x$$

To find the gradient at a given point, we substitute the value of x. For example, the gradient of the tangent at the point $(3, 9)$ is $2(3) = 6$.

The process of finding the gradient of the tangent at any point on the curve of $f(x) = x^2$ produced a new function $f'(x) = 2x$, which is called the derivative of the function f. If we write $y = f(x)$, we denote the derivative as dy/dx. The process of computing the derivative of a function is called differentiation.

The derivative of a function has the following interpretations.

1. The function $f'(x)$ represents the instantaneous rate of change of $y = f(x)$ with respect to x.

2. The function $f'(x)$ represents the gradient of the graph of $f(x)$ at any point.

3. The function $f'(x)$ represents the gradient of the tangent at any point on the graph of $f(x)$.

Differentiation from first principles

Finding the derivative of a function using the definition is called differentiation from first principles.

Examples

Find the derivative of each of the following functions.

1. $f(x) = x^3$

Let Δx and Δy represent small increment in x and y respectively.

$$\frac{\Delta y}{\Delta x} = \frac{f(x+\Delta\Delta x) - f(x)}{\Delta x}$$

$$= \frac{(x+\Delta x)^3 - x^3}{\Delta x}$$

$$= \frac{3x^2\Delta x + 3x(\Delta x)^2 + (\Delta x)^3}{\Delta x}$$

$$= 3x^2 + 3x\Delta x + (\Delta x)^2$$

So, $\lim\limits_{\Delta x \to 0} \dfrac{\Delta y}{\Delta x} = \dfrac{dy}{dx} = 3x^2$

2. $f(x) = 2x$

$$\frac{\Delta y}{\Delta x} = \frac{f(x+\Delta x) - f(x)}{\Delta x}$$

$$= \frac{2(x+\Delta x) - 2x}{\Delta x}$$

$$= \frac{2\Delta x}{\Delta x}$$

$$= 2$$

So, $\lim\limits_{\Delta x \to 0} \dfrac{\Delta y}{\Delta x} = \dfrac{dy}{dx} = 2$

3. $f(x) = -3$

$$\frac{\Delta y}{\Delta x} = \frac{f(x+\Delta x) - f(x)}{\Delta x}$$

$$= \frac{-3 - (-3)}{\Delta x}$$

$$= 0$$

So, $\lim\limits_{\Delta x \to 0} \dfrac{\Delta y}{\Delta x} = \dfrac{dy}{dx} = 0$

Differentiation

The notation dy/dx is read "the derivative of y with respect to x". A variable other than x may be used as the independent variable. For example, if $y = f(t)$, then the derivative of y with respect to t could be written $f'(t)$ or dy/dt.

Exercise 15.1(a)

Find the derivative of each of the following functions.

1. $f(x) = x$
2. $f(x) = 5x$
3. $f(x) = 3x^2$
4. $f(x) = -2x^2$
5. $f(x) = 2$
6. $f(x) = -½x^2$
7. $f(x) = ⅔x^3$
8. $f(x) = -3x^2 + 3$
9. $f(x) = 1 - 2x^3$

Rules for finding Derivatives

Constant Rule

If $f(x) = k$, where k is any real number, then $f'(x) = 0$.

The derivative of a constant is 0.

Powers Rule

For any real number n,

$$\frac{d}{dx}(x^n) = nx^{n-1}.$$

Examples

Find the derivative of each function.

1. $f(x) = x^5$

$$f'(x) = 5x^4$$

2. $y = x^{-3}$

$$\frac{dy}{dx} = -3x^{-4}$$

CHAPTER 15 Differentiation and Integration

Exercise 15.1(b)

Find the derivative of each of the following functions.

1. $y = -x$
2. $y = 8$
3. $y = x^4$
4. $y = x^6$
5. $y = -x^{-3}$
6. $y = x^{-1}$
7. $y = x^{-8}$
8. $y = x^{-3/4}$
9. $y = -x^{1/2}$
10. $y = x^{-3/2}$
11. $y = -x^{-2/3}$
12. $y = -x^{1/3}$

Constant Times a Function

If f is a differentiable function and k is a real number, then

$$\frac{d}{dx}[kf(x)] = kf'(x)$$

The derivative of constant times a function is the constant times the derivative of the function.

Examples

Find the derivative of each of the following functions.

1. $f(x) = -3x^2$

$$f(x) = -3x^2$$
$$f'(x) = -3 \cdot 2x$$
$$= -6x$$

2. $y = \frac{1}{2}x^6$

$$y = \frac{1}{2}x^6$$
$$\frac{dy}{dx} = \frac{1}{2} \cdot 6x^5$$
$$= 3x^5$$

Differentiation

Exercise 15.1(c)

Find the derivative of each of the following functions.

1. $y = -2x$
2. $y = 3x^4$
3. $y = -2x^5$
4. $y = \frac{1}{4}x^8$
5. $y = \frac{2}{3}x^6$
6. $y = -\frac{2}{5}x^{-5}$
7. $y = 2x^{-3}$
8. $y = -4x^{-3}$
9. $y = 3x^{-2}$
10. $y = 3x^{1/3}$
11. $y = -2x^{3/2}$
12. $y = 6x^{1/2}$
13. $y = -6x^{-2/3}$
14. $y = 8x^{-3/4}$
15. $y = 9x^{-2/3}$

Sum or Difference Rule

If f and g are differentiable functions, then

$$\frac{d}{dx}[f(x) \pm g(x)] = f'(x) \pm g'(x)$$

The derivative of a sum or difference of functions is the sum or difference of the derivatives.

Examples

Find the derivative of each of the following functions.

1. $f(x) = 2x^{-3} + 5$

$$f(x) = 2x^{-3} + 5$$
$$f'(x) = -6x^{-4}$$

2. $y = x^3 + 3x^2 - 2x$

$$y = x^3 + 3x^2 - 2x$$
$$\frac{dy}{dx} = 3x^2 + 6x - 2$$

Exercise 15.1(d)

Find the derivative of each of the following functions.

Differentiation and Integration

1. $y = 2x^3 + 3x^4$
2. $y = 3x^5 - 2x$
3. $y = 3 - t^4 + t^5$
4. $y = 7 - 2t^{-1}$
5. $y = 4t^3 + 3t$
6. $y = 5x^4 - 3x^5$
7. $y = 5 - x^{-3}$
8. $y = 2x^3 - 3x^{-2}$
9. $s = 3t^4 - 4t + 2$
10. $s = 2t^2 - 3t$
11. $y = 6x^2 + 5x + 2$
12. $s = \frac{1}{3}t^3 - \frac{1}{2}t^{-2}$
13. $y = 3x^{1/3} - 2x^{-1/2}$
14. $s = 4t^{-3/4} + \frac{1}{3}t^3$
15. $y = 2 - 6x^{-2/3}$
16. $y = 1 + \frac{1}{x^3}$
17. $y = 3x - \frac{1}{\sqrt{x}} + \frac{1}{x}$
18. $y = \frac{5}{x^2} - \frac{1}{\sqrt{x^3}} + 2$
19. $y = t^2 + \frac{4}{t^3}$
20. $y = 2x(3 - x)$
21. $y = 3(x - 2)^2$
22. $y = \frac{3x^2 - 2x^3}{3x}$
23. $y = \frac{3x^2 + 2x - 1}{x}$
24. $s = \frac{(t+3)(2t-1)}{t^2}$

The Product Rule

If f and g are differentiable, then

$$\frac{d}{dx}[f(x)g(x)] = f(x) \cdot g'(x) + g(x) \cdot f'(x)$$

The derivative of a product of two functions is the first function times the derivative of the second function plus the second function times the derivative of the first.

Examples

Find the derivative of each of the following functions.

1. $f(x) = x^3(2x - 3)$

$$f(x) = x^3(2x - 3)$$
$$f'(x) = x^3 \cdot 2 + (2x - 3) \cdot 3x^2$$
$$= 2x^3 + 3x^2(2x - 3)$$
$$= 2x^3 + 6x^3 - 9x^2$$
$$= 8x^3 - 9x^2$$

2. $y = (3x^2 + 2x)(4x + 3)$

$$y = (3x^2 + 2x)(4x + 3)$$

$$\frac{dy}{dx} = (3x^2 + 2x) \cdot 4 + (4x + 3) \cdot (6x + 2)$$

$$= 12x^2 + 8x + 24x^2 + 26x + 6$$

$$= 36x^2 + 34x + 6$$

Exercise 15.1(e)

Find the derivative of each of the following functions.

1. $y = x^3(4x^2 - 3)$
2. $y = (2x - 3)(x + 4)$
3. $y = 3x^2(2x + 1)$
4. $y = (2x + 1)(3x - 2)$
5. $y = x(2x^3 - 3x^2)$
6. $y = x(3 - 2x)(2 + x)$
7. $y = \sqrt{x}(2x + 3)$
8. $y = (1 - x^2)(1 + 2x^2)$
9. $y = \sqrt[3]{x}(2 + x)$
10. $y = 3x^{-2}(2x - 3)$
11. $y = (x^4 - 1)(x^3 + 1)$
12. $y = (2x - 5)(3x^2 + 2)$

The Quotient Rule

If f and g are differentiable functions and $g'(0) \neq 0$, then

$$\frac{d}{dx}\left[\frac{f(x)}{g(x)}\right] = \frac{g(x) \cdot f'(x) - f(x) \cdot g'(x)}{[g(x)]^2}$$

The derivative of a quotient is the denominator times the derivative of the numerator minus the numerator times the derivative of the denominator, all divided by the square of the denominator.

Example

Find the derivative of $y = \frac{2x}{3x-1}$.

$$y = \frac{2x}{3x-1}$$

$$\frac{dy}{dx} = \frac{(3x-1) \cdot 2 - 2x \cdot 3}{(3x-1)^2}$$

$$= \frac{6x - 2 - 6x}{(3x-1)^2}$$

$$= -\frac{2}{(3x-1)^2}$$

Exercise 15.1(f)

Find the derivative of each of the following functions.

1. $y = \frac{x}{x+1}$
2. $y = \frac{x+1}{x-1}$
3. $y = \frac{1}{x^2-1}$
4. $y = \frac{2}{3x^2-1}$
5. $y = \frac{3x}{x^2-2}$
6. $y = \frac{3x^2-5}{2x-7}$
7. $y = \frac{1-\sqrt{x}}{1+\sqrt{x}}$
8. $y = \frac{1+x^2}{1-x^2}$
9. $y = \frac{x^2}{x+1}$
10. $y = \frac{x^2}{x^2-1}$
11. $y = \frac{x^2}{2x+3}$
12. $y = \frac{\sqrt{x}}{\sqrt{x}-1}$

The Chain Rule

Many functions are composite functions of simpler functions. For instance, if f and g are functions such that $f(x) = x^3$ and $g(x) = 3x + 2$ and $y = (3x + 2)^3$, then $y = f[g(x)]$. Suppose y is a function of u and u is itself a function of x, then $y = f(u)$, and

$$\frac{dy}{dx} = \frac{dy}{du} \times \frac{du}{dx}.$$

This formula is called the chain rule.

Example

Find the derivative of $y = (3x - 2)^4$.

$$y = (3x - 2)^4$$

Let $u = 3x - 2$

then $y = u^4$

$$\frac{dy}{du} = 4u^3$$

and $\frac{du}{dx} = 3$

By the chain rule

$$\frac{dy}{dx} = \frac{dy}{du} \cdot \frac{du}{dx}$$

$$= 4u^3 \cdot 3$$

$$= 12u^3$$

$$= 12(3x - 2)^3$$

Exercise 15.1(g)

Find the derivative of each of the following functions.

1. $y = (x - 5)^3$
2. $y = (3x - 2)^5$
3. $y = (1 - 2x^2)^6$
4. $y = (x^2 + 1)^4$
5. $y = (x^3 - 2)^7$
6. $y = (2x + 1)^{-3}$
7. $y = (3 - 2x)^{-4}$
8. $y = (4x + 3)^{-2}$
9. $y = (1 - 2x^2)^{-1}$
10. $y = (x^2 - 1)^{1/2}$
11. $y = (3x + 2)^{-2/3}$
12. $y = (2x^2 - 3)^{3/4}$
13. $y = \frac{1}{(3x^2 + 2x)^2}$
14. $y = \frac{1}{\sqrt[3]{1-x}}$
15. $y = \frac{1}{\sqrt{x^2 - 1}}$
16. $y = \frac{1}{\sqrt{2 + x^3}}$

An alternative formula of the chain rule is stated below.

$$\frac{d}{dx}[f(x)]^n = n[f(x)]^{n-1} \cdot f'(x)$$

This formula helps you write in one step the derivative of a function of a function.

Example

Find the derivative of $y = (3 - 2x)^5$

$$y = (3 - 2x)^5$$

$$\frac{dy}{dx} = 5(3 - 2x)^4 \cdot -2$$

$$= -10(3 - 2x)^4$$

Exercise 15.1(h)

Find the derivative of each of the following functions.

1. $y = (x + 3)^4$
2. $y = (3x - 2)^3$
3. $y = (2x^2 + 3)^5$
4. $y = (x^2 + 1)^{1/2}$
5. $y = (2x^3 - 1)^{-1/3}$
6. $y = (3 + x^2)^{-2}$
7. $y = (1 - x^2)^3$
8. $y = (3x^2 - 1)^{-3}$
9. $y = (1 - x^2)^6$
10. $y = \frac{1}{(x^2-3)^4}$
11. $y = \frac{1}{\sqrt{2x^2+3}}$
12. $y = \frac{1}{\sqrt[3]{x^3+3x}}$

Derivatives of Trigonometric Functions

The derivative of four trigonometric functions, measured in radian, is summarized in the table below.

y	dy/dx
$\sin x$	$\cos x$
$\cos x$	$-\sin x$
$\tan x$	$\sec^2 x$
$\cot x$	$-\coses^2 x$

Differentiation

We can use the chain rule to find derivatives of other trigonometric functions, as shown in the following examples.

Examples

Find the derivative of each of the following functions.

1. $y = \sin 3x$

$$y = \sin 3x$$

$$\text{Let } u = 3x$$

$$\text{then } y = \sin u$$

$$\frac{dy}{du} = \cos u$$

$$\text{and } \frac{du}{dx} = 3$$

The chain rule gives

$$\frac{dy}{dx} = \frac{dy}{du} \times \frac{du}{dx}$$

$$= \cos u \times 3$$

$$= 3 \cos 3x$$

2. $y = x \tan x + 1$

Using the product rule and differentiating term by term, we have

$$\frac{dy}{dx} = \tan x + x \sec^2 x$$

3. $y = \cos^2 x$

$$\text{Let } u = \cos x$$

$$\text{then } y = u^2$$

$$\frac{dy}{du} = 2u \text{ and } \frac{du}{dx} = -\sin x$$

The chain rule gives

CHAPTER 15 Differentiation and Integration

$$\frac{dy}{dx} = \frac{dy}{du} \times \frac{du}{dx}$$

$$= 2u \times -\sin x$$

$$= -2u \sin x$$

$$= -2 \cos x \sin x$$

$$= -\sin 2x$$

You can obtain the same result by using the product rule as follows.

$$y = (\cos x)(\cos x)$$

$$\frac{dy}{dx} = \cos x \cdot -\sin x + (-\sin x \cdot \cos x)$$

$$= -\sin x \cos x - \sin x \cos x$$

$$= -2 \sin x \cos x$$

$$= -\sin 2x$$

Exercise 15.1(i)

1. Find the derivative of each of the following functions.

(a) $y = \cos 2x$
(b) $y = \sin 5x$
(c) $y = \tan 3x$
(d) $y = 2\cos 3x$
(e) $y = -\frac{1}{2}\sin 6x$
(f) $y = 4\cot 2x$
(g) $y = \cos(3x - 2)$
(h) $y = \cot(5 - 2x)$
(i) $y = \sin(2x + 3)$
(j) $y = \sin x^2$
(k) $y = 6\cos\frac{1}{2}x$
(l) $y = \tan 3x^2$
(m) $y = \tan(2x - 5)$
(n) $y = \cot\frac{1}{3}x^2$
(o) $y = \tan(x^2 + 1)$

2. Find the derivative of each of each of the following functions.

(a) $y = \cos^2 2x$
(b) $y = 3\sin^2 x$
(c) $y = \tan^2 2x$
(f) $y = \sin^2 3x$

3. Find the derivative of each of the following functions.

   (a) $y = x \cos x$
   (b) $y = x \sin 2x$
   (c) $y = x^2 \sin x$
   (d) $y = x^3 \tan x$

4. Find the derivative of each of the following functions.

   (a) $y = 2x \sin \frac{1}{2}x$
   (b) $y = x \cos 2x$
   (c) $y = x^2 \cos x$
   (d) $y = x^2 \tan 2x$
   (e) $y = \sin x \cos x$
   (f) $y = x^3 \cot 2x$
   (g) $y = \cos x \tan x$
   (h) $y = \frac{\sin x}{x}$
   (i) $y = \frac{\cos 2x}{x}$

5. Find the derivative of each of the following functions.

   (a) $y = 2 \sin x + 3x$
   (b) $y = 1 - \cos 2x$
   (c) $y = x^2 + 3 \tan x$
   (d) $y = 2 \cos x + 3$
   (e) $y = 2 \sin x + x^2$
   (f) $y = 2x^3 - 3 \cos x$
   (g) $y = 2x^2 + 5 \tan x$
   (h) $y = 3 \sin 2x - \cos x$
   (i) $y = x \tan x - 3x^2$
   (j) $y = x^2 \sin x - 1$
   (k) $y = x^2 \tan x + \frac{1}{x}$
   (l) $y = x^2 \sin x - \frac{1}{x^2}$

Derivatives of Logarithmic Functions

The logarithmic function with base e is the natural logarithmic function, and is defined by the symbol $\ln x$. It can be shown that if $y = \ln x$, then

$$\frac{dy}{dx} = \frac{1}{x}$$

Examples

Find the derivative of each of the following functions.

1. $y = \ln 3x$

$$y = \ln 3x$$

Let $u = 3x$

then $y = \ln u$

$$\frac{dy}{du} = \frac{1}{u} \quad \text{and} \quad \frac{du}{dx} = 3$$

The chain rule gives

$$\frac{dy}{dx} = \frac{dy}{du} \times \frac{du}{dx}$$

$$= \frac{1}{u} \times 3$$

$$= \frac{3}{3x}$$

$$= \frac{1}{x}$$

You can obtain the same result as follows.

$$y = \ln 3x$$

$$= \ln 3 + \ln x$$

$$\frac{dy}{dx} = 0 + \frac{1}{x}$$

$$= \frac{1}{x}$$

2. $y = \ln(2x - 3)$

$$\frac{dy}{dx} = \frac{1}{2x-3} \times 2$$

$$= \frac{2}{2x-3}$$

03. $y = \ln x^2$

$$\frac{dy}{dx} = \frac{1}{x^2} \times 2x$$

$$= \frac{2}{x}$$

Differentiation

Exercise 15.1(j)

Find the derivative of each of the following functions.

1. $y = \ln \frac{1}{2}x$
2. $y = \ln 4x$
3. $y = 2 \ln \frac{1}{3}x$
4. $y = \ln(3x + 2)$
5. $y = \ln(1 - 2x)$
6. $y = \frac{1}{4}\ln(4x + 3)$
7. $y = \ln x^3$
8. $y = \ln 2\sqrt[3]{x}$
9. $\ln \left(\frac{1}{\sqrt{x}}\right)$
10. $y = \ln \left(\frac{1}{x^3}\right)$
11. $y = \ln \left(\frac{2}{x^2}\right)$
12. $\ln \left(\frac{1}{\sqrt[3]{x^2}}\right)$
13. $y = x \ln x^2$
14. $y = x^3 \ln x$
15. $y = x \ln(2x - 1)$
16. $y = x^2 \ln \frac{1}{2}x$
17. $y = x \ln 3x^2$
18. $y = x^3 \ln(3x + 2)$

Derivatives of Exponential Functions

The exponential function $f(x) = e^x$ and the natural logarithmic function $g(x) = \ln x$ are inverse functions of each other. The exponential function can be written in an equivalent logarithmic form.

If $y = e^x$, then $x = \ln y$.

$$\text{So, } \frac{dx}{dy} = \frac{1}{y}$$

$$\text{Now, } \frac{dy}{dx} = \frac{1}{dx/dy}$$

$$= y$$

$$= e^x$$

The derivative of e^x is e^x.

CHAPTER 15   Differentiation and Integration

Examples

Find the derivative of each of the following functions.

1. $y = e^{2x}$

$$y = e^{2x}$$

$$\text{Let } u = 2x$$

$$\text{then } y = e^u$$

$$\frac{dy}{du} = e^u$$

$$\text{and } \frac{du}{dx} = 2$$

The chain rule gives

$$\frac{dy}{dx} = \frac{dy}{du} \times \frac{du}{dx}$$

$$= e^2 \times 2$$

$$= 2e^{2x}$$

2. $y = e^{3x-2}$

$$y = e^{3x-2}$$

$$\frac{dy}{dx} = e^{3x-2} \times 3$$

$$= 3e^{3x-2}$$

3. $y = e^{x^2}$

$$y = e^{x^2}$$

$$\frac{dy}{dx} = e^{x^2} \times 2x = 2xe^{x^2}$$

Exercise 15.1(k)

Find the derivative of each of the following functions.

1. $y = e^{3x}$ 2. $y = e^{\frac{1}{2}x}$ 3. $y = e^{-\frac{3}{4}x}$

4. $y = e^{-2x}$ 5. $y = e^{3x+2}$ 6. $y = e^{1-2x}$

7. $y = e^{\frac{1}{3}x+2}$ 8. $y = e^{-\frac{2}{3}x+5}$ 9. $y = e^{3x^2}$

10. $y = e^{1-x^3}$ 11. $y = e^{\sqrt{x}}$ 12. $y = e^{\sqrt[3]{x^2}}$

13. $y = xe^{2x^2}$ 14. $y = x^2 e^{\sqrt{x}}$ 15. $y = xe^{-\frac{1}{3}x^2}$

16. $y = (x+1)e^x$ 17. $y = xe^{1-x^2}$ 18. $y = (x^2-1)e^{x^3}$

Tangents and Normals

The tangent to a curve at a point is the straight line that just touches the curve at that point. A normal to a curve at a point is the straight line through the point at right angles to the tangent at the point.

Recall that, the derivative of the function $y = f(x)$ represents the gradient of the tangent to the graph of the function at a point. Remember that, the gradients of two perpendicular lines are negative reciprocal of each other.

We can find the equation of a tangent and the equation of a normal by using the formula $y - y_1 = m(x - x_1)$.

Examples

1. Find the equation of the tangent to the curve $y = 3x^2 - 2x - 5$ at the point where $x = 2$.

   We have $\frac{dy}{dx} = 6x - 2$.

   So, the gradient of the tangent at $x = 2$ is

   $$\frac{dy}{dx} = 6(2) - 2 = 10$$

   Next, we find the y-coordinate of the point where $x = 2$.

   $$y = 3(2^2) - 2(2) - 5 = 3$$

   Then the equation of the tangent is

   $$y - 3 = 10(x - 2)$$

   $$y = 10x - 17$$

2. Find the equation of the normal to the curve $y = x^2 - 2x$ at the point where $x = 3$.

Now,

$$\frac{dy}{dx} = 2x - 2$$

The gradient of the tangent is

$$\frac{dy}{dx} = 2(3) - 2 = 4.$$

So, the gradient of the normal is $-¼$.

The y-coordinate of the point where $x = 3$ is

$$y = 3^2 - 2(3) = 3$$

Then, the equation of the normal is

$$y - 3 = -\frac{1}{4}(x - 3)$$

$$4y - 12 = -x + 3$$

$$x + 4y - 15 = 0$$

Exercise 15.1(I)

1. Find the gradient of the tangent to each of the following curves at the given points.

   (a) $y = x^2 - x + 1$, $(2, 3)$
   (b) $y = 3 + 4x - 2x^2$, $(3, -3)$
   (c) $y = 3x^2 - 2$, $(0, -2)$
   (d) $y = x^3 - 3x$, $(2, 2)$
   (e) $y = x^3 + 2x^2 + 1$, $(-2, 1)$
   (f) $y = 3x - x^3$, $(2, 1)$
   (g) $y = x^4 - 3x^2 + 2x - 1$, $(1, -1)$
   (h) $y = x^2 - 8/x^3$, $(-2, 5)$

2. Find the equation of the tangent to each of the following curves at the given points.

   (a) $y = 2x^2 + 3$, $(-1, 5)$
   (b) $y = 2x^2 - 3x - 5$, $(3, 4)$
   (c) $y = 1 + 2x^2 - x^3$, $(2, 1)$
   (d) $y = x^3(2 - x^2)$, $(1, 1)$

(e) $y = x^3 + 2x^2 - 3x - 1$, $(-2, 5)$ (f) $y = 3 - 4/x^2$, $(2, 2)$

3. Find the equation of the normal to each of the following curve at the given points.

   (a) $y = 3x + x^2$, $(1, 4)$ (b) $y = 2x^2 - 3$, $(-2, 5)$

   (c) $y = 2 + 3x^2 - x^3$, $(2, 6)$ (d) $y = x^3 + 2x$, $(1, 3)$

   (e) $y = 3x^3 + 2x^2 - 5x + 3$, $(-2, -3)$ (f) $y = (3x - 2)^6$, $(1, 1)$

4. At what point on the curve $y = 3x^2 + 4x - 5$ is the tangent parallel to the line $y = 7x - 19$.

5. At what point on the curve $y = 2x^2 - 3x + 1$ is the tangent parallel to the line $y = 5x - 7$.

6. At what point on the curve $y = x^3 - 2x^2$ is the normal parallel to the line $x + 4y - 17 = 0$.

7. At what point on the curve $y = 2x^2 - 3x + 4$ is the normal parallel to the line $3x - y + 14 = 0$.

Implicit Differentiation

Often functions are defined in the form $y = f(x)$, with y given explicitly in terms of x, or as an explicit function of x. For example, $y = 2x + 3$, $y = x^3 + 3x^2 + x$ and $y = 2 - x^2$ are all explicit functions of x.

The equation $2xy + 3x^2 = 4$ can be expressed as an explicit function of x by solving for y. This gives $y = (4 - 3x^2)/2x$. For some equations in x and y, it would be either tedious or impossible to solve for y. In such cases, y is said to be given implicitly in terms of x. An example of an implicit function is $y^3 + 6y^2 + 4xy + 1 = 0$.

We can find the derivative of implicit functions by a process called implicit differentiation. To differentiate implicit functions, we regard the whole expression as a function of x and differentiate both sides of the equation.

Because y is expressed as a function of x, the chain rule is used whenever the function y is differentiated. For example,

$$\frac{d}{dx}(y^2) = \frac{d}{dy}(y^2) \cdot \frac{dy}{dx} = 2y\frac{dy}{dx}$$

CHAPTER 15 Differentiation and Integration

and $\frac{d}{dx}(y^3) = \frac{d}{dy}(y^3) \cdot \frac{dy}{dx} = 3y^2 \frac{dy}{dx}$.

Example

Find the derivative of $y^3 + 2xy - x^3 = 5$.

Differentiate both sides of the equation with respect to x.

$$y^3 + 2xy - x^3 = 5$$

$$3y^2 \frac{dy}{dx} + 2x \frac{dy}{dx} + 2y - 3x^2 = 0$$

$$(3y^2 + 2x)\frac{dy}{dx} = 3x^2 - 2y$$

$$\frac{dy}{dx} = \frac{3x^2 - 2y}{3y^2 + 2x}$$

Exercise 15.1(m)

1. Find dy/dx by implicit differentiation for the following function.

   (a) $y^2 - x^2 = 8$
   (b) $2x - y^2 = 5$
   (c) $x^2 - 4xy + y^2 = 0$
   (d) $y^2 - 2xy = x^2$
   (e) $x^3 - y^3 = 2$
   (f) $3x^2 - 4xy = 0$
   (g) $x^2 - y^2 + 6y + 8x = 0$
   (h) $x^2 + 3xy = y^2$
   (i) $x^2 + y^2 + 2xy - 3y + 2x = 7$
   (j) $x^3 y^2 = 1$

2. Find the gradient to the following curves at the given points.

   (a) $xy = 6$, $(3, 2)$
   (b) $x^2 - y^2 = -8$, $(1, 3)$
   (c) $x^3 + y^3 = 7$, $(-1, 2)$
   (d) $3x^3 + 2y^2 = 14$, $(2, 1)$
   (e) $x^2 + 3xy - y^2 = 9$, $(2, 1)$
   (f) $x^3 - 2xy + y^3 = 0$, $(1, 1)$

3. Find the equation of the tangent and normal to each of the following curves at the given points.

   (a) $x^2 - xy + y^2 = 7$, $(3, 2)$
   (b) $x^2 + y^2 + 3xy - 11 = 0$, $(1, 2)$
   (c) $y^2 - 3x + 2y + x^2 = 1$, $(2, 3)$
   (d) $x^2 y + x^3 + 2y = 2$, $(2, -1)$

Second Order Derivative

If a function f has a derivative f', then the derivative of f', if it exists, is the second derivative of f, written f''.

The second derivative of $y = f(x)$ can also be written

$$\frac{d}{dx}\left(\frac{dy}{dx}\right) = \frac{d^2y}{dx^2}.$$

d^2y/dx^2 is read "d two y by $d\,x$ squared ".

Examples

Find the second derivative of the following functions.

1. $y = 3x^4 - 2x^3$

$$y = 3x^4 - 2x^3$$

$$\frac{dy}{dx} = 12x^3 - 6x^2$$

$$\frac{d^2y}{dx^2} = 36x^2 - 12x$$

2. $f(x) = (x^2 + 1)^3$

$$f(x) = (x^2 + 1)^3$$

$$f'(x) = 3(x^2 + 1)^2 \cdot 2x$$

$$= 6x(x^2 + 1)^2$$

$$f''(x) = 6x \cdot 2(x^2 + 1) \cdot 2x + (x^2 + 1)^2 \cdot 6$$

$$= 24x^2(x^2 + 1) + 6(x^2 + 1)^2$$

$$= 6(x^2 + 1)(5x^2 + 1)$$

Exercise 15.1(n)

Find the second derivative of each of the following functions.

1. $y = 3x^2$
2. $y = 4x$
3. $y = x^3 - 3x^2$
4. $y = x^4 + x^2 - 2$

5. $y = 3x^2 + 2x$
6. $y = 3 - 2x + x^2 - x^3$
7. $y = x^{-4}$
8. $y = 2x^{-3}$
9. $y = 2 + 3x^{-1}$
10. $y = x^{-3} - 2x$
11. $y = 3x + \frac{2}{x^2}$
12. $y = 1 - \frac{4}{x^3}$
13. $y = (1 + x^3)^2$
14. $y = (\sqrt{x} + 1)^2$
15. $y = (2x - 3)^3$
16. $y = \frac{x}{x+1}$
17. $y = \frac{1+x}{1-x}$
18. $y = \frac{1+x^2}{1-x^2}$

Local Maxima and Local Minima

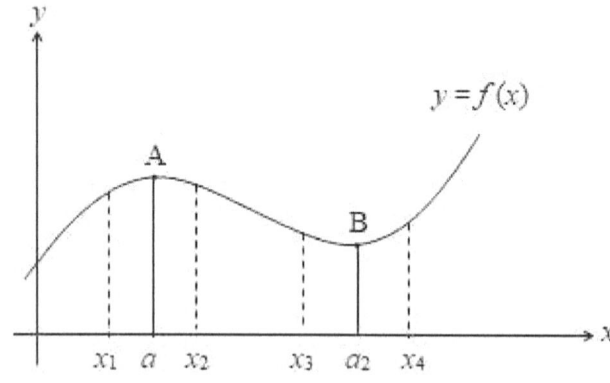

Figure 15.4

Figure 15.4 shows the graph of a function $y = f(x)$. At the point A, i.e. at $x = a_1$ a maximum value of y occurs since at A, the y-value is greater than the y-values on either side of it and close to it. The greatest value of the function in the interval $x_1 < x < x_2$, is $f(a_1)$, called the local maximum (or relative maximum).

Similarly, at B, a minimum value of y occurs, since the y-value at the point B is less than the y-values on either side of it and close to it. The least value of the function in the interval $x_3 < x < x_4$ is $f(a_2)$, called the local minimum (or relative minimum).

Point A and B are called stationary points (or turning points) on the graph or stationary values of y.

The point $(a_1, f(a_1))$ is a maximum point and the point $(a_2, f(a_2))$ is a minimum point.

A function may have more than one relative maximum or relative minimum. The plural of relative maximum is relative maxima, and the plural of relative minimum is relative minima.

First Derivative Test

Remember that, the derivative of a function at a point gives the gradient of the tangent to the function at that point. In Section 8.4, you learned that a line with a positive gradient rises from left to right and a line with a negative gradient falls from left to right.

The graphs shown in Figure 15.5 and Figure 15.6 illustrate the first derivative test.

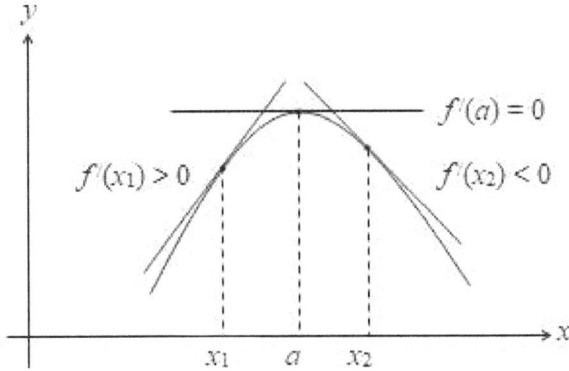

Figure 15.5

As shown in Figure 15.5, the gradient of the curve at the point $x = x_1$ is positive and the gradient at the point $x = x_2$ is negative. The function has a maximum value at $x = a$. At this point the tangent is horizontal and has gradient 0. On the left side of the maximum point, the tangent has positive gradient, indicating that the function is increasing. On the right, the tangent has negative gradient, indicating that the function is decreasing. Notice that the sign of the gradient changes from positive through zero to negative, as you move from left to right.

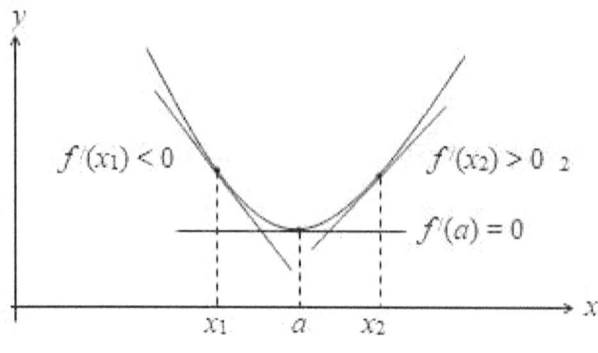

Figure 15.6

In Figure 15.6, the gradient of the curve at the point $x = x_1$ is negative and the gradient at the point $x = x_2$ is positive. The function has a minimum value at $x = a$. At this point the tangent is horizontal and has gradient 0. On the left side of the minimum point, the tangent has negative gradient, indicating that the function is decreasing. On the right, the tangent has positive gradient, indicating that the function is increasing. Notice that the sign of the gradient changes from negative through zero to positive, as you move from left to right.

Locating Stationary Points

When a function is given as an equation, we can use differentiation to determine the possible stationary points on the graph of the function.

A function f has a stationary point at $x = a$ when $f'(a) = 0$. We can determine the x-coordinate of a stationary point by solving the equation $f'(x) = 0$. To identify whether a stationary point is a maximum or a minimum, we may use the first derivative test.

Examples

Find the coordinates of all stationary points and determine the nature of the stationary point.

1. $y = x^2 - 6x + 5$

   The derivative is

$$\frac{dy}{dx} = 2x - 6$$

Setting the derivative equal to 0 gives

$$2x - 6 = 0$$

$$2x = 6$$

$$x = 3$$

The *y*-coordinate of the stationary point is

$$y = 3^2 - 6(3) + 5 = -4$$

The point (3, −4) is a stationary point.

To determine whether this point is a maximum or a minimum we investigate the sign of the gradient on either side of the point.

We take two numbers, one on each side of 3. On the left, we a take a number less than and close to 3, and on the right we take a number greater than and close to 3.

We will choose 2.5 and 3.5. Evaluate *dy*/*dx* at 2.5 and 3.5.

$$\frac{dy}{dx} = 2(2.5) - 6 = -1 < 0$$

and

$$\frac{dy}{dx} = 2(3.5) - 6 = 1 > 0$$

The working can be summarized in a table, as shown below.

We use L for left and R for right.

Value of x	L	3	R
Sign of dy/dx	−	0	+
	\	−	/

The gradient is negative on the left and positive on the right. By the first derivative test the function has minimum at $x = 3$.

Thus, (3, −4) is a minimum point.

252 CHAPTER 15 Differentiation and Integration

2. $y = 3 + 2x - x^2$

$$\frac{dy}{dx} = 2 - 2x$$

At a stationary point $dy/dx = 0$.

$$2 - 2x = 0$$

$$x = 1$$

The y-coordinate of the stationary point is

$$y = 3 + 2 - 1 = 4$$

The point (1, 4) is a stationary point.

Value of x	L	1	R
Sign of dy/dx	+	0	−
	/	−	\

Thus, the point (1, 4) is a maximum.

3. $y = x^3 + 6x^2 + 9x - 8$

$$\frac{dy}{dx} = 3x^2 + 12x + 9$$

At a stationary point $dy/dx = 0$

$$3x^2 + 12x + 9 = 0$$

$$x^2 + 4x + 3 = 0$$

$$(x + 1)(x + 3) = 0$$

$$x + 1 = 0 \quad \text{or} \quad x + 3 = 0$$

$$x = -1 \qquad\qquad x = -3$$

When $x = -1$,

$$y = -1 + 6 - 9 - 8 = -12$$

The point $(-1, -12)$ is a stationary point.

Value of x	L	-1	R
Sign of dy/dx	$-$	0	$+$
	\	$-$	/

The point $(-1, -12)$ is a minimum point.

When $x = -3$,

$$y = (-3)^3 + 6(-3)^2 + 9(-3) - 8 = -8$$

The point $(-3, -8)$ is a stationary point.

Value of x	L	-3	R
Sign of dy/dx	$+$	0	$-$
	/	$-$	\

The point $(-3, -8)$ is a maximum point.

Exercise 15.1(o)

Find the coordinates of the stationary points of the following functions. In each case investigate the nature of the stationary point.

1. $y = x^2 - 4x + 3$
2. $y = 5 + 2x - x^2$
3. $y = 2x^2 - x^3$
4. $y = 3x^2 - x^3$
5. $y = x^3 - 3x - 7$
6. $y = x^4 - 4x^3$
7. $y = 2x^3 + 3x^2 - 12x + 6$
8. $y = 2x^3 - x^2 - 8x + 3$
9. $y = 15x - x^2 - \frac{1}{3}x^3$
10. $y = \frac{1}{3}x^3 - \frac{5}{2}x^2 + 6x$
11. $y = x^2(x - 2)$
12. $y = x(2x - 3)(x - 4)$

Second Derivative Test

Another way to determine the nature of a stationary point is to use the second derivative test.

If f is a function such that $f'(a) = 0$, then

1. $(a, f(a))$ is a maximum point if $f''(a)$ is negative.

2. $(a, f(a))$ is a minimum point if $f''(a)$ is positive.

Note: If the second derivative is 0, then use the first derivative test.

Examples

Find the coordinates of all stationary points and then use the second derivative test to determine the nature of the stationary point.

1. $y = 2x^2 - 8x + 3$

$$\frac{dy}{dx} = 4x - 8$$

At a stationary point $dy/dx = 0$.

$$4x - 8 = 0$$

$$x = 2$$

When $x = 2$,

$$y = 2(2)^2 - 8(2) + 3 = -5$$

The point $(2, -5)$ is a stationary point.

To determine whether this point is a maximum or minimum we find the second derivative.

The second derivative is $d^2y/dx^2 = 4$, which is positive indicating a minimum.

The point $(2, -5)$ is a minimum point.

2. $y = 5 + 12x - 2x^2$

$$\frac{dy}{dx} = 12 - 4x$$

At a stationary point $dy/dx = 0$.

$$12 - 4x = 0$$

$$x = 3$$

When $x = 3$.

$$y = 5 + 12(3) - 2(3)^2 = 23$$

The point (3, 23) is a stationary point.

Now, $d^2y/dx^2 = -4$, which is negative indicating a maximum.

The point (3, 23) is a maximum point.

3. $y = 3x^3 - 3x^2 + 2$

$$\frac{dy}{dx} = 9x^2 - 6x$$

At a stationary point $dy/dx = 0$.

$$9x^2 - 6x = 0$$

$$3x(3x - 2) = 0$$

$$3x = 0 \quad \text{or} \quad 3x - 2 = 0$$

$$x = 0 \qquad\qquad x = \frac{2}{3}$$

When $x = 0$, $y = 2$.

The point (0, 2) is a stationary point.

The second derivative is $d^2y/dx^2 = 18x - 6$.

Evaluating d^2y/dx^2 at $x = 0$ gives

$$\frac{d^2y}{dx^2} = 18(0) - 6 = -6,$$

which is negative indicating a maximum.

The point (0, 2) is a maximum point.

When $x = 2/3$

$$y = 3\left(\frac{2}{3}\right)^3 - 3\left(\frac{2}{3}\right)^2 + 2 = \frac{14}{9}$$

The point (2/3, 14/9) is a stationary point.

Evaluating d^2y/dx^2 at $x = 2/3$ gives

$$\frac{d^2y}{dx^2} = 18\left(\frac{2}{3}\right) - 6 = 6,$$

which is positive indicating a minimum.

The point (2/3, 14/9) is a minimum point.

Exercise 15.1(p)

Find the coordinates of all stationary points and then use the second derivative test to determine the nature of the stationary point.

1. $y = x^2 - 6x$
2. $y = x - 4x^2$
3. $y = x^2 - 4x + 7$
4. $y = 5 + 6x - x^2$
5. $y = 2x^3 - x^2 - 8x + 3$
6. $y = x + \frac{1}{x}$
7. $y = 2x^3 - 11x^2 - 12x - 5$
8. $y = x - \frac{4}{x^2}$
9. $y = 8x + 5x^2$
10. $y = x^3 - 3x + 7$
11. $y = x^3 - 6x^2 + 9x + 8$
12. $y = 3x - x^3$
13. $y = x^3 - 3x^2 + 4$
14. $y = 4x - 3x^3$
15. $y = x^2(2 - x)$
16. $y = 4x + \frac{1}{x}$
17. $y = 2x^2 + \frac{1}{2x}$
18. $y = x - \frac{4}{x^2}$

Application of Maxima and Minima

The steps used in finding stationary points and investigating the nature of the points can be used to solve practical problems, as shown in the following examples.

Examples

Differentiation

1. Thin metal is used to make cylindrical cans which are to hold 1,200 cm³ of fruit juice. What should be the radius of the can if they are to use the least amount of metal?

Finding the least amount of metal require minimizing the surface area of the can. Let the radius and height of the can be r cm and h cm respectively.

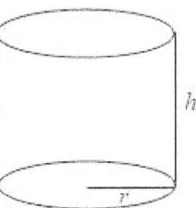

The surface area of the top and the bottom is $2\pi r^2$ and the area of the side is $2\pi rh$. Thus, the surface area, A, is given by

$$A = 2\pi r^2 + 2\pi rh$$

The right side of the equation involves two variables. We cannot differentiate until we have removed one of the variables. We can do this by using the information about the volume, $V = \pi r^2 h$, of the can.

Since the volume of the can is 1,200 cm³, we have

$$\pi r^2 h = 1200$$

Solve this for h.

$$h = \frac{1200}{\pi r^2}$$

Substitute this expression for h into the equation for A.

$$A = 2\pi r^2 + 2\pi r \left(\frac{1200}{\pi r^2}\right)$$

$$= 2\pi r^2 + \frac{2400}{r}$$

Now, find dA/dr and solve $dA/dr = 0$ for r.

$$\frac{dA}{dr} = 4\pi r - \frac{2400}{r^2}$$

$$4\pi r - \frac{2400}{r^2} = 0$$

$$r^3 = \frac{600}{\pi}$$

$$r = \sqrt[3]{\frac{600}{\pi}}$$

$$r \approx 5.8$$

Now, we investigate the nature of the stationary point at r = 5.8.

The second derivative is

$$\frac{d^2A}{dr^2} = 4\pi + \frac{4800}{r^3}.$$

Since r is positive the second derivative is always positive, so there is a minimum at $r = 5.8$.

The can will have least amount of metal when $r = 5.8$ cm.

2. An open box is to be made by cutting a square from each corner of 8 cm by 5 cm sheet of metal and then folding up the sides. Find the dimensions of the box with the maximum volume.

Let x represents the length of a side of the square that is cut from each corner. Then the length of the box is $8 - 2x$ and the width is $5 - 2x$, as shown in the diagram. The depth of the box is x.

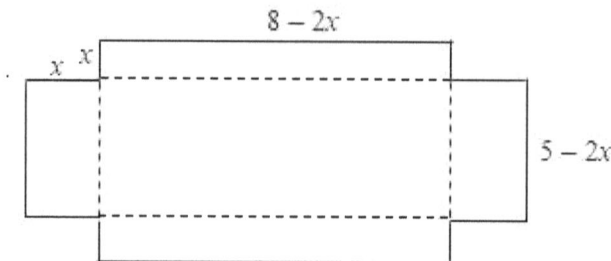

The volume of the box is given by length × width × depth. Thus,

$$V = x(8 - 2x)(5 - 2x)$$

$$= 40x - 26x^2 + 4x^3$$

Now, find dV/dx and solve $dV/dx = 0$ for x.

$$\frac{dV}{dx} = 40 - 52x + 12x^2$$

$$40 - 52x + 12x^2 = 0$$

$$(3x - 10)(x - 1) = 0$$

$$3x - 10 = 0 \quad \text{or} \quad x - 1 = 0$$

$$x = \frac{10}{3} \qquad x = 1$$

Now, $d^2y/dx^2 = -52 + 24x$

When $x = 10/3$,

$$\frac{d^2V}{dx^2} = -52 + 24\left(\frac{10}{3}\right) = 28$$

which indicate a minimum.

When $x = 1$,

$$\frac{d^2V}{dx^2} = -52 + 24 = -28$$

which indicate a maximum.

Evaluate the volume at $x = 1$.

$$V = 40 - 26 + 4 = 18$$

The maximum volume is 18 cm^3, and this is achieved if we cut squares of side 1 cm out of each corner.

The dimensions of the maximum box are 6 cm long by 3 cm wide by 1 deep.

The examples suggest the following guidelines.

260 CHAPTER 15   Differentiation and Integration

1. Draw a diagram and label it.

2. Determine what the variables are and how they are related.

3. Decide what quantity needs to be maximized or minimized.

4. Write an expression for the quantity to be maximized or minimized in terms of only one variable.

5. Determine the minimum and maximum values.

6. Answer the question that is asked.

Exercise 15.1(q)

1. A cuboid box is to have a volume of 576 cm^3. The length is twice the breadth. Find the height, if the surface area is to be as small as possible.

2. A metal basket is made in the form of a cylinder with an open top. Its height is h cm and its radius r cm. It is to contain 3,000 cm^3. What is the radius, if it is to consist of the smallest possible amount of metal?

3. A cardboard box is to be made in the form of a cuboid with square cross-section. Its volume must be 8000 cm^3. Let x be the side of the square, and z cm be the length. What value of x will use the least cardboard?

4. A closed cylindrical can is made of a tin plate. If the volume of the can is 64 cm^3, find the radius of the can with the least possible surface area.

5. A manufacturer of tin cans wishes to produce a closed cylindrical can of volume 2 litres. Find the dimensions of the can with the least possible surface area.

6. A metal can is made in the form of a cylinder with an open top. Its height is h cm and its radius r cm. It is to contain 10 cm^3 of liquid. What is the radius, if it is to consist of the smallest possible amount of metal?

7. A right- angled triangle has a hypotenuse of 9 m. Find the maximum area, as the other two sides vary.

8. Find the maximum possible area of a rectangle whose perimeter is 32 cm. What are the dimensions?

9. A three-sided fence is to be built by a farmer next to a straight section of a river, which forms the fourth side of a rectangular field. If there is 200 m of fencing available, find the maximum enclosed area and the dimensions of the corresponding enclosure.

10. A rectangular fence is to be built by a farmer. If the enclosed area is to equal 900 m^2, find the minimum perimeter and the dimensions of the corresponding enclosure.

11. An open box is made by cutting a square region from each corner of a sheet of metal 8 cm by 5 cm and folding up the sides. Find the length of the side of each square region that must be cut so that the volume of the box will be maximum.

12. An open box is made by cutting a square region from each corner of a sheet of metal 12 cm square and folding up the sides. Find the length of the side of each square region that must be cut out so that the volume of the box will be maximum.

13. The bottom of a tank of height h cm is a square of side x m and the tank is open at the top. It is designed to hold 4 m^3 of liquid. Express in terms of x the total area of the bottom and the four sides of the tank. Find the value of x for which the area is minimum.

14. An open tank is to be constructed with a square horizontal base and vertical sides. The capacity of the tank is to be 400 m^3. The cost of the material for the sides is £5 per square metre and the base is £3 per square metre. Find the minimum cost of the material, and give the corresponding dimensions of the tank.

15. A fence must be built in a large field to enclose a rectangular area of 10,000 m^2. One side of the area is bounded by an existing fence; no fence is needed there. Material for the fence cost £ 2 per metre for the ends and £1.50 per metre for the side opposite the existing fence. Find the cost of the least expensive fence.

16. A closed box with a square base is to have a volume of 6750 cm^3. The material for the top and bottom of the box cost £3 per square centimetre, while the material for the sides cost £1.50 per square centimetre. Find the dimensions of the box that will lead to the minimum total cost. What is the minimum total cost?

17. A company manufactures cylindrical metal containers with volume of 16 m^2. The top and bottom of each container is made of a material that cost £2 per square metre, while the material for the side cost £1 per square metre. Find the radius, height and cost of the least expensive container.

Curve Sketching

The method of finding stationary points and determining their nature allows us to sketch graphs of functions.

Sketching graphs of functions may be done with the following steps.

1. Find the intercept on the x- and y- axes.

2. Find all possible stationary points.

3. Find the behaviour of the function (if necessary) as x becomes very large or very small.

4. Plot the intercepts, the stationary points and other points as needed.

5. Finally, connect the points with a smooth curve.

Example

Sketch the curve of $y = x^3 - 3x^2$.

Begin by finding the x and y intercepts.

To find the y-intercept put $x = 0$, giving $y = 0$.

The y-intercept is (0, 0).

To find the x-intercept put $y = 0$.

$$x^3 - 3x^2 = 0$$

$$x^2(x - 3) = 0$$

giving $x = 0$ or $x = 3$.

The x-intercepts are (0, 0) and (3, 0).

Now, find the stationary points and determine their nature.

$$\frac{dy}{dx} = 3x^2 - 6x$$

For maximum or minimum, $dy/dx = 0$.

$$3x^2 - 6x = 0$$

$$3x(x - 2) = 0$$

giving $x = 0$ or $x = 2$.

When $x = 0$, $y = 0$.

When $x = 2$,

$$y = 2^3 - 3(2^2) = -4.$$

The points $(0, 0)$ and $(2, -4)$ are stationary points.

Next, we determine the nature of the stationary points.

$$\frac{d^2y}{dx^2} = 6x - 6$$

When $x = 0$,

$$\frac{d^2y}{dx^2} = 6(0) - 6 = -6$$

which indicates a maximum.

Thus, $(0, 0)$ is a maximum point.

When $x = 2$,

$$\frac{d^2y}{dx^2} = 6(2) - 6 = 6$$

which indicates a minimum.

Thus $(2, -4)$ is a minimum point.

Finally, we examine the behaviour of the function as x increases or decreases.

As $x \to \pm\infty$, $y \approx x^3$.

Thus, as $x \to +\infty$, $y \to +\infty$ and as $x \to -\infty$, $y \to -\infty$.

From these results, we obtain a graph as shown below.

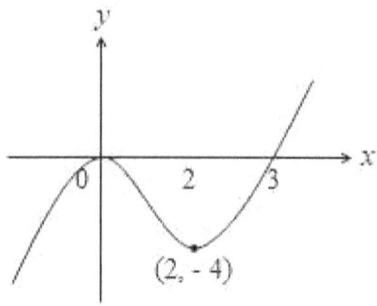

Exercise 15.1(r)

Sketch the graph of each of the following functions.

1. $y = x^2 - x - 6$
2. $y = 3 + 2x - x^2$
3. $y = x^3 - 12x$
4. $y = x^3 - 3x$
5. $y = 3x^2 - x^3$
6. $y = x^3 - 6x^2$
7. $y = x^3 - 2x^2 + x$
8. $y = x^2(x - 1)$
9. $y = x^3 - 6x^2 + 9x$
10. $y = x^4 - 2x^3$
11. $y = 4x^5 - 5x^4$
12. $y = 4x^3 + 3x^4$

Small Increments and Approximations

For any function $y = f(x)$, a small increment Δx in x causes a change Δy in y. The rate of change of y with respect to x is $\Delta y/\Delta x$. For small values of Δx and Δy,

$$\frac{\Delta y}{\Delta x} \approx \frac{dy}{dx}$$

Multiplying both sides by Δx (provided $\Delta x \neq 0$) gives

$$\Delta y \approx \frac{dy}{dx} \cdot \Delta x$$

Examples

1. If the side of a cube is increased from 12 cm to 12.01 cm, what is the approximate increase in volume?

   Let the volume be V and the side x.

   Thus, $V = x^3$

   The change in length $\Delta x = 0.01$.

   $$\frac{dV}{dx} = 3x^2$$

   $$\Delta V = \frac{dV}{dx} \cdot \Delta x$$

   $$= 3x^2 \cdot 0.01$$

   $$= 0.03x^2$$

   When $x = 12$

   $$\Delta V = 0.03 \cdot 12^2 = 4.32$$

   Thus, the approximate increase in volume is 4.32 cm.

2. Find an approximation for $\sqrt[3]{125.1}$

   Let $y = \sqrt[3]{x}$

   If we let $x = 125$, then $\Delta x = 0.1$.

   $$\frac{dy}{dx} = \frac{1}{3}x^{-2/3} = \frac{1}{3\sqrt[3]{x^2}}$$

   $$\Delta y = \frac{dy}{dx} \cdot \Delta x$$

   $$= \frac{1}{3\sqrt[3]{x^2}} \cdot 0.1$$

   Substituting 125 for x and 0.1 for Δx give

   $$\Delta y = \frac{0.1}{3\sqrt[3]{125^2}} = 0.0013$$

   Thus, $\sqrt[3]{125.1} \approx 5.0013$

Exercise 15.1(s)

1. Find the approximate increase in area of a square when its side changes from 10 cm to 10.1 cm.

2. A cube has side 8 cm. Find the approximate change in its volume if its side changes by 0.02 cm.

3. The radius of a sphere is measured as 6.5 cm, with possible error of 0.03 cm. What is the possible error in the volume?

4. The side of a square is measured with a possible error of 3 %, what is the approximate percentage error in the area?

5. The percentage error when measuring the area of a circle was 5 %. What was the approximate percentage error in its radius?

6. What is the error in area of a circle and its circumference if its radius is 0.2 % greater than its correct radius of 1 m?

7. A 1 ½ % error is made in measuring the radius of a sphere. Find the percentage error in surface area.

8. An error of 2 % is made in measuring the radius of a sphere. What are the resulting errors in the calculation of its surface area and volume?

9. The height of a cylinder is 6 cm and its radius is 3 cm. Find the approximate increase in volume when the radius increases to 3.02 cm.

10. The volume of a sphere increases by 2 %. Find the corresponding percentage increase in surface area.

Find an approximation for each the following.

11. $\sqrt{5.3}$

12. $\sqrt[3]{27.1}$

13. $\sqrt[5]{32.01}$

14. $\sqrt{17.2}$

15. $\sqrt[4]{81.02}$

16. $\sqrt[3]{29}$

17. $\sqrt{147}$

18. $\sqrt[5]{36}$

19. $\sqrt[3]{1003}$

Rate of Change

Often, each quantity in an equation changes with time (or some other variable). For example, if a cylindrical container collects water from a tap, the volume, V, and depth, h of water increases with time t. Notice that the volume of water also increases as the depth of water increases. By the chain rule, we could obtain an equation which connects the rate of change of volume and depth. Since V is a function of h, and both V and h are functions of t, then by the chain rule

$$\frac{dV}{dt} = \frac{dV}{dh} \cdot \frac{dh}{dt}$$

Using this equation, we could calculate either the rate of change of volume or depth if one of them is given.

Solving related rates problems usually involve three steps.

1. Find the functional relationship between the variables in the problem.

2. Use the chain rule to write an equation connecting the rate of change of the variables.

3. Substitute the given values, and then solve for the derivative giving the unknown rate of change.

Examples

1. The side of a cube is increasing at 5 cm s^{-1}, what is the rate of increase of the volume when $x = 3$ cm.

    Let the volume be V and the side x.

    Thus, $V = x^3$

    $$\frac{dV}{dx} = 3x^2$$

    By the chain rule

    $$\frac{dV}{dt} = \frac{dV}{dx} \cdot \frac{dx}{dt}$$

    $$= 3x^2 \cdot 5$$

    $$= 15x^2$$

    When $x = 3$

    $$\frac{dV}{dt} = 15 \cdot 3^2 = 135$$

The rate of increase of the volume is 135 cm^3 s^{-1}.

The same result is obtained as follows.

$$V = x^3$$

Differentiate both sides with respect to t.

$$\frac{d}{dt}(V) = \frac{d}{dt}(x^3)$$

$$\frac{dV}{dt} = 3x^2 \cdot \frac{dx}{dt}$$

$$= 3x^2 \cdot 5$$

$$= 15x^2$$

When $x = 3$

$$\frac{dV}{dt} = 15 \cdot 3^2$$

$$= 135 \text{ cm}^3 \text{ s}^{-1}$$

2. Air is pumped into a spherical balloon at 12 cm^3 s^{-1}. How fast is the radius increasing when it is 3 cm?

Let the volume be V and radius r. Thus,

$$V = \frac{4}{3}\pi r^3$$

By the chain rule

$$\frac{dV}{dt} = \frac{dV}{dr} \cdot \frac{dr}{dt}$$

$$12 = 4\pi r^2 \cdot \frac{dr}{dt}$$

$$\frac{dr}{dt} = \frac{12}{4\pi r^2}$$

$$= \frac{1}{3\pi}$$

The rate of increase of the radius is $1/3\pi$ cm s^{-1}.

Exercise 15.1(t)

1. The side of a square is increasing at 0.8 cm s^{-1}. At what rate is the area increasing at a time when the side is 10 cm?

2. The radius of a sphere is increasing at 0.2 cm s^{-1}. At what rate is the surface area increasing when the radius is 3 cm? At what rate is the volume increasing?

3. The area of a circle is increasing at 12 cm^2 s^{-1}. At what rate is the radius increasing, when it is 20 cm?

4. Water is poured into a cone of vertical angle 90° at 10 cm^3 s^{-1}. When the height of water is 15 cm, at what rate is it increasing?

5. The volume of a cube is decreasing at 6 cm^3 s^{-1}. When the side is 4 cm, what is the rate of decrease of (a) the side, and (b) the surface area?

6. A pump is inflating a spherical balloon. If the radius at a certain instant is 5 cm and it is increasing at a rate of 10 cm s^{-1}, at what rate is the pump working?

7. If air is pumped into a balloon at the rate of 0.50 m^3 s^{-1} at what rate will the radius be increasing when it is 5 m?

8. A spherical balloon is losing air at a rate of 18 m^3 s^{-1}. When its radius is 3 m, at what rate is it diminishing?

9. Liquid is dropping through a conical funnel at a rate of 5 cm^3 s^{-1}. When the depth of liquid in the funnel is x cm, its volume is $1/3\pi x^3$. Find the rate at which the level of liquid is falling when $x = 10$.

10. A funnel is made in the shape of a right circular cone of height 8 cm and base radius 6 cm. Water drains from the vertex of the funnel at the rate of 1.8 cm^3 s^{-1}. Find the rate at which the water level is dropping when the water has receded 4 cm from the top.

15.2 Integration

Indefinite Integrals

Integration is the reverse of finding a derivative. That is, to integrate a function $f(x)$ is to find another function $F(x)$ whose derivative is $f(x)$. The function $F(x)$ is generally denoted by $\int f(x) \, dx$, and is called the indefinite integral of $f(x)$, with respect to x. For example, if we differentiate x^3 we obtain $3x^2$. So, $\int 3x^2 \, dx = x^3 + c$,

where c is a constant, called the constant of integration (or arbitrary constant).

The process of computing an integral is called integration.

Rules of Integration

Power Rule

For any real number $n \neq -1$,

$$\int x^n \, dx = \frac{x^{n+1}}{n+1} + c,$$

where c is a constant.

The integration of a power of x is found by increasing the index by 1, and divide by the new index.

Examples

Find the following integrals.

1. $\int x^7 \, dx$

$$\int x^7 \, dx = \frac{x^{7+1}}{7+1} + c$$

$$= \frac{1}{8} x^8 + c$$

2. $\int x^{-5} \, dx$

$$\int x^{-5} \, dx = \frac{x^{-5+1}}{-5+1} + c$$

$$= -\frac{1}{4} x^{-4} + c$$

3. $\int x^{1/2} \, dx$

$$\int x^{1/2} \, dx = \frac{x^{\frac{1}{2}+1}}{\frac{1}{2}+1} + c$$

$$= \frac{2}{3} x^{3/2} + c$$

Exercise 15.2(a)

Find the following integrals.

1. $\int 5 \, dx$
2. $\int \frac{1}{3} \, dx$
3. $\int x^2 \, dx$
4. $\int x^4 \, dx$
5. $\int x^9 \, dx$
6. $\int x^{11} \, dx$
7. $\int x^{-3} \, dx$
8. $\int x^{-2} \, dx$
9. $\int x^{-5} \, dx$
10. $\int x^{1/3} \, dx$
11. $\int x^{3/4} \, dx$
12. $\int x^{3/2} \, dx$
13. $\int x^{-1/4} \, dx$
14. $\int x^{-2/3} \, dx$
15. $\int x^{-4/3} \, dx$
16. $\int \sqrt[5]{x} \, dx$
17. $\int \sqrt[3]{x^2} \, dx$
18. $\int \sqrt[4]{x^5} \, dx$
19. $\int \frac{1}{x^4} \, dx$
20. $\int \frac{1}{x^7} \, dx$
21. $\int \frac{1}{x^6} \, dx$
22. $\int \frac{1}{\sqrt[5]{x^4}} \, dx$
23. $\int \frac{1}{\sqrt{x^3}} \, dx$
24. $\int \frac{1}{\sqrt[4]{x^3}} \, dx$

Constant Multiple

$$\int k \cdot f(x) \, dx = k \int f(x) \, dx,$$

for any real number k.

To integrate a function multiplied by a constant, integrate the function and then multiply by the constant.

Examples

Find the following integrals.

1. $\int 2x^3 \, dx$

$$\int 2x^3 \, dx = 2 \int x^3 \, dx$$
$$= 2 \cdot \frac{1}{4} x^4 + c$$
$$= \frac{1}{2} x^4 + c$$

2. $\int 6x^{-4} \, dx$

$$\int 6x^{-4} \, dx = 6 \int x^{-4} \, dx$$
$$= 6 \cdot -\frac{1}{3} x^{-3} + c$$
$$= -2x^{-3} + c$$

Exercise 15.2(b)

Find the following integrals.

1. $\int 6x \, dx$
2. $\int 4x^5 \, dx$
3. $\int 8x^3 \, dx$
4. $\int \frac{3}{4} x^2 \, dx$
5. $\int \frac{5}{4} x^4 \, dx$
6. $\int \frac{2}{3} x^7 \, dx$
7. $\int \frac{8}{x^3} \, dx$
8. $\int \frac{4}{x^2} \, dx$
9. $\int \frac{3}{x-2} \, dx$
10. $\int \frac{1}{3x^2} \, dx$
11. $\int \frac{2}{3x^4} \, dx$
12. $\int \frac{3}{\sqrt{x}} \, dx$

Sum or Difference Rule

$$\int [f(x) \pm g(x)] \, dx = \int f(x) \, dx \pm \int g(x) \, dx$$

To integrate the sum of functions, integrate term by term.

Examples

Find the following integrals.

1. $\int (x^3 - x^2 + 2) \, dx$

$$\int (x^3 - x^2 + 2) \, dx = \int x^3 \, dx - \int x^2 \, dx + \int 2 \, dx$$
$$= \frac{1}{4} x^4 - \frac{1}{3} x^3 + 2x + c$$

2. $\int(x+3)(x+2)\,dx$

$$\int(x+3)(x+2)\,dx = \int(x^2+5x+6)\,dx$$
$$= \frac{1}{3}x^3 + \frac{5}{2}x^2 + 6x + c$$

3. $\int \frac{x^4+x}{x^3}\,dx$

$$\int \frac{x^4+x}{x^3}\,dx = \int(x+x^{-2})\,dx$$
$$= \frac{1}{2}x^2 - x^{-1} + c$$
$$= \frac{1}{2}x^2 - \frac{1}{x} + c$$

Exercise 15.2(c)

Find the following integrals.

1. $\int(2x-3)\,dx$
2. $\int(3x+5)\,dx$
3. $\int(x^2-8)\,dx$
4. $\int(x^4+x^3)\,dx$
5. $\int(2x^2+x^3)\,dx$
6. $\int(2x^3-5x^4)\,dx$
7. $\int(3x^2-2x+1)\,dx$
8. $\int(3-2x+4x^2)\,dx$
9. $\int(1-2x+6x^2)\,dx$
10. $\int(5x^4-3x^2+2x)\,dx$
11. $\int x(x^2-2)\,dx$
12. $\int x(4x^2+3x)\,dx$
13. $\int x^2(2+3x^2)\,dx$
14. $\int x(x^4-3x)\,dx$
15. $\int 2x^2(2x-x^2)\,dx$
16. $\int 3x^2(x^2+1)\,dx$
17. $\int(x+1)(x-3)\,dx$
18. $\int(2x+1)(x+3)\,dx$
19. $\int(3-2x)^2\,dx$
20. $\int x(x-2)^2\,dx$
21. $\int(x^2-3x)\sqrt{x}\,dx$
22. $\int(x^2+2x)\sqrt[3]{x}\,dx$
23. $\int \frac{x^3+1}{x^2}\,dx$
24. $\int \frac{5x-3x^2}{x}\,dx$

274 CHAPTER 15   Differentiation and Integration

25. $\int \frac{x^2+2x}{x^5} dx$

26. $\int \frac{(x-2)^2}{\sqrt[3]{x}} dx$

27. $\int \frac{\sqrt{x}-5}{x^2} dx$

28. $\int \frac{1+x^2}{\sqrt{x}} dx$

29. $\int \frac{4-3x^2+2x^3}{x^2} dx$

30. $\int \frac{(1+x^2)(1-2x^2)}{x^2} dx$

Definite Integrals

If $F(x)$ is any integral of the function $f(x)$, then the definite integral of f, denoted by

$$\int_a^b f(x)dx$$

is given by

$$\int_a^b f(x)dx = F(b) - F(a)$$

The b above the integral sign is called the upper limit of integration, and the a is called the lower limit of integration.

Examples

Evaluate each definite integral.

1. $\int_1^2 x^3 \, dx$

$$\int_1^2 x^3 \, dx = \left[\frac{1}{4}x^4\right]_1^2$$

$$= \left(\frac{1}{4} \times 2^4\right) - \left(\frac{1}{4} \times 1^4\right)$$

$$= 4 - \frac{1}{4}$$

$$= \frac{15}{4}$$

The symbol $[f(x)]_a^b$ is used to represent $F(b) - F(a)$.

2. $\int_0^2 (3x^2 - 4x^3) \, dx$

$$\int_0^2 (3x^2 - 4x^3)\, dx = [x^3 - x^4]_0^2$$
$$= (2^3 - 2^4) - 0$$
$$= -8$$

Exercise 15.2(d)

Evaluate the following integrals.

1. $\int_{-1}^{2} x\, dx$
2. $\int_{1}^{3} x^2\, dx$
3. $\int_{2}^{4} x^3\, dx$

4. $\int_{0}^{3} 3x^2\, dx$
5. $\int_{-2}^{-1} x^4\, dx$
6. $\int_{-1}^{1} \frac{3}{2}x^2\, dx$

7. $\int_{-2}^{2}(x^2 + x)\, dx$
8. $\int_{0}^{2}(4 - x^2)\, dx$
9. $\int_{2}^{4}(3x^2 - 2x)\, dx$

10. $\int_{1}^{2}(x^3 - x)\, dx$
11. $\int_{-1}^{2}(3 - x^2)\, dx$
12. $\int_{1}^{3}(x^2 - 4x + 3)\, dx$

13. $\int_{1}^{2} x(1 + 4x)\, dx$
14. $\int_{1}^{4} x(x + 3)\, dx$
15. $\int_{-1}^{2}(x^3 + 2)^2\, dx$

16. $\int_{-1}^{1}(2x + 1)^3\, dx$
17. $\int_{-1}^{1} x(x + 1)(x + 2)\, dx$
18. $\int_{0}^{1}(x + 2)\sqrt{x}\, dx$

19. $\int_{1}^{2} \frac{4}{x^3}\, dx$
20. $\int_{1}^{9} \frac{1}{\sqrt{x}}\, dx$
21. $\int_{4}^{9} \frac{1}{\sqrt{x^3}}\, dx$

22. $\int_{1}^{2} \frac{1}{x^3}\, dx$
23. $\int_{2}^{3} \frac{2}{3x^2}\, dx$
24. $\int_{1}^{4} \sqrt{x}\, dx$

25. $\int_{1}^{3} \frac{2x^2+1}{x^2}\, dx$
26. $\int_{1}^{2} \frac{x^3+2x}{x^3}\, dx$
27. $\int_{-1}^{2} \frac{x^3+2x^2-3x}{x}\, dx$

28. $\int_{1}^{4} \frac{x^2-1}{\sqrt{x}}\, dx$
29. $\int_{1}^{2}\left(1 - \frac{8}{x^3}\right) dx$
30. $\int_{1}^{9}\left(\sqrt{x} + \frac{1}{\sqrt{x}}\right) dx$

Integration by Substitution

Occasionally, to find the integral $\int f(x)\, dx$, we make the substitution $x = g(x)$ and use the following rule

$$\int f(x)\, dx = \int f(g(u)) g'(u)\, du$$

Examples

1. Use substitution to find the following integrals.

(a) $\int (3x-2)^3 \, dx$

Let $u = 3x - 2$, then $du = 3dx$

Now substitute u for $3x - 2$ and ⅓ du for dx.

$$\int (3x-2)^3 \, dx = \int u^3 \cdot \frac{du}{3}$$

$$= \frac{1}{3} \int u^3 \, du$$

$$= \frac{1}{3}\left(\frac{1}{4}u^4\right) + c$$

$$= \frac{1}{12}u^4 + c$$

Now replace u with $3x - 2$.

$$\int (3x-2)^3 \, dx = \frac{1}{12}(3x-2)^4 + c$$

(b) $\int \frac{15}{(5x-3)^2} \, dx$

Let $u = 5x - 3$, then $du = 5dx$.

$$\int \frac{15}{(5x-3)^2} \, dx = \int \frac{15}{u^2} \cdot \frac{du}{5}$$

$$= 3 \int u^{-2} \, du$$

$$= -3u^{-1} + c$$

$$= -\frac{3}{u} + c$$

$$= -\frac{3}{5x-3} + c$$

2. Use substitution to find

$$\int_{-1}^{1} x^2(1-x^3) \, dx$$

Let $u = 1 - x^3$, then $du = -3x^2 \, dx$.

Because we have change the variable we must also change the limits of integration to correspond to the new variable.

When $x = 1$, $u = 1 - 1 = 0$

and when $x = -1$, $u = 1 + 1 = 2$. Thus,

$$\int_{-1}^{1} x^2(1 - x^3)\, dx = \int_{2}^{0} x^2 \cdot \frac{u}{-3x^2}\, du$$

$$= -\frac{1}{3} \int_{2}^{0} u\, du$$

$$= \left[-\frac{1}{6} u^2\right]_{2}^{0}$$

$$= \frac{2}{3}$$

Note: you can use the original limits when you substitute $1 - x^3$ for u.

Exercise 15.2(e)

Use substitution to find the following integrals.

1. $\int (3x + 2)^3\, dx$
2. $\int (2x - 1)^{-2}\, dx$
3. $\int \sqrt[3]{x - 2}\, dx$
4. $\int x(x^2 + 3)^2\, dx$
5. $\int x(x^2 + 1)^4\, dx$
6. $\int x^2(2x^3 + 3)^4\, dx$
7. $\int x\sqrt{1 - x^2}\, dx$
8. $\int x^2\sqrt{x^3 - 2}\, dx$
9. $\int x\sqrt{4x^2 - 5}\, dx$
10. $\int x^2\sqrt{x^3 + 4}\, dx$
11. $\int x^3\sqrt{2 - x^4}\, dx$
12. $\int x^2\sqrt{2x^3 - 3}\, dx$
13. $\int \frac{1}{(2x-5)^4}\, dx$
14. $\int \frac{1}{\sqrt[3]{3x+1}}\, dx$
15. $\int \frac{2x}{\sqrt{1+3x^2}}\, dx$
16. $\int \frac{3x}{(x^2-1)^4}\, dx$

Evaluate the following definite integrals.

17. $\int_0^1 (1 - x)^4\, dx$
18. $\int_2^7 \sqrt{x + 2}\, dx$
19. $\int_0^2 x^2\sqrt{x^3 + 1}\, dx$
20. $\int_{-1}^{2} \frac{1}{(2x+1)^2}\, dx$

278 CHAPTER 15   Differentiation and Integration

21. $\int_1^2 \frac{1}{(3x-2)^3} dx$ 22. $\int_0^1 \frac{x^3}{(x^4+1)^3} dx$

Integration of Trigonometric Functions

The integral of the trigonometric functions given in the following table can be used to find integrals of other trigonometric functions.

y	$\int y\, dx$
$\sin x$	$-\cos x$
$\cos x$	$\sin x$
$\sec^2 x$	$\tan x$

Examples

Find the following integrals.

1. $\int \sec^2 15x\, dx$

$$\text{Let } u = 15x, \text{ then } du = 15\, dx.$$

$$\int \sec^2 15x\, dx = \frac{1}{15}\int \sec^2 u\, du$$

$$= \frac{1}{15}\tan u + c$$

Replacing u with $15x$ gives

$$\int \sec^2 15x\, dx = \frac{1}{15}\tan 15x + c$$

2. $\int \cos(3x - 2)\, dx$

$$\text{Let } u = 3x - 2, \text{ then } 3\, dx.$$

$$\int \cos(3x - 2)\, dx = \frac{1}{3}\int \cos u\, du$$

$$= \frac{1}{3}\sin u + c$$

$$= \frac{1}{3}\sin(3x - 2) + c$$

3. $\int x \sin x^2\, dx$

Let $u = x^2$, then $du = 2x\, dx$.

$$\int x \sin x^2\, dx = \frac{1}{2} \int \sin u\, du$$

$$= -\frac{1}{2} \cos u + c$$

$$= -\frac{1}{2} \cos x^2 + c$$

Exercise 15.2(f)

Find the following integrals.

1. $\int \cos 2x\, dx$
2. $\int \sin 3x\, dx$
3. $\int 2 \sec^2 4x\, dx$
4. $\int 3 \sin 2x\, dx$
5. $\int 3 \cos 6x\, dx$
6. $\int -6 \sin 4x\, dx$
7. $\int \frac{3}{4} \sin 3x\, dx$
8. $\int \sec^2 \frac{1}{2} x\, dx$
9. $\int \cos(2x + 1)\, dx$
10. $\int 3 \sin(3 - 2x)\, dx$
11. $\int \cos(5x + 2)\, dx$
12. $\int \sec^2(3x - 2)\, dx$
13. $\int \sin(3 - \frac{1}{2} x)\, dx$
14. $\int 6 \sec^2(\frac{3}{2} x + 2)\, dx$
15. $\int \cos(2 - 3x)\, dx$
16. $\int x^2 \sin x^3\, dx$
17. $\int 3x \cos x^2\, dx$
18. $\int 2x^2 \sec^2 x^3\, dx$
19. $\int \sin^2 x \cos x\, dx$
20. $\int \sin x \cos^3 x\, dx$

Evaluate the following definite integrals.

21. $\int_0^{\pi/2} \sin x\, dx$
22. $\int_0^{\pi/4} \cos x\, dx$
23. $\int_0^{\pi/3} \sec^2 x\, dx$
24. $\int_{\pi/2}^{2\pi/3} \sin x\, dx$

Integrals of Exponential Functions

The integral of the exponential function e^x is itself. That is,

$$\int e^x \, dx = e^x + c$$

Examples

Find the following integrals.

1. $\int e^{2x} \, dx$

   Let $u = 2x$, then $du = 2 \, dx$.

   $$\int e^{2x} \, dx = \frac{1}{2} \int e^u \, du$$

   $$= \frac{1}{2} e^u + c$$

   $$= \frac{1}{2} e^{2x} + c$$

2. $\int e^{2-3x} \, dx$

   Let $u = 2 - 3x$, then $du = -3 \, dx$.

   $$\int e^{2-3x} \, dx = -\frac{1}{3} \int e^u \, du$$

   $$= -\frac{1}{3} e^u + c$$

   $$= -\frac{1}{3} e^{2-3x} + c$$

3. $\int x^3 e^{x^4} \, dx$

   Let $u = x^4$, then $du = 4x^3 \, dx$.

   $$\int x^3 e^{x^4} \, dx = \frac{1}{4} \int e^u \, du$$

   $$= \frac{1}{4} e^u + c$$

   $$= \frac{1}{4} e^{x^4} + c$$

Exercise 15.2(g)

Find the following integrals.

1. $\int e^{3x}\,dx$
2. $\int e^{-2x}\,dx$
3. $\int e^{-\frac{3}{4}x}\,dx$
4. $\int e^{\frac{2}{3}x}\,dx$
5. $\int e^{2x+3}\,dx$
6. $\int e^{-3x+2}\,dx$
7. $\int 8e^{3-4x}\,dx$
8. $\int e^{\frac{1}{2}x+3}\,dx$
9. $\int 2e^{4-\frac{1}{3}x}\,dx$
10. $\int 3e^{4x}\,dx$
11. $\int -2e^{-4x}\,dx$
12. $\int 6e^{-2x}\,dx$
13. $\int xe^{x^2}\,dx$
14. $\int 2x^2 e^{x^3}\,dx$
15. $\int xe^{3-x^2}\,dx$

15.3 Area under Curves

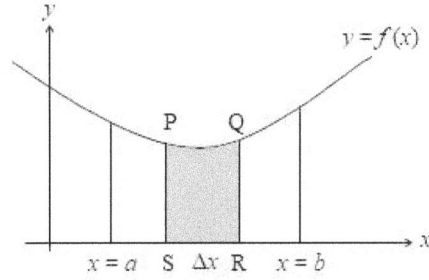

Figure 15.7

Consider the area bounded by the curve $y = f(x)$, the x-axis and the ordinates at $x = a$ and $x = b$. Let A represents the area under the curve up to the point P on the curve with coordinates (x, y). If we move the point P to the point Q with coordinates $(x + \Delta x, y + \Delta y)$, the right-hand boundary of A moves from PS to QR and the area is increased by a small area ΔA. The area of PQRS can be approximated by the area of a rectangle whose height is y and width Δx. Thus,

$$\Delta A \approx y \Delta x$$

Dividing this by Δx gives

$$\frac{\Delta A}{\Delta x} \approx y$$

As $\Delta x \to 0$, the limit of $\Delta A/\Delta x$ is

$$\frac{dA}{dx} = y.$$

It follows that $A = \int y \, dx$.

This represents the area under the curve up to the point P.

The area under the curve between $x = a$ and $x = b$ is the difference between area up to $x = b$ and the area up to $x = a$, written briefly as

$$A = \int_a^b y \, dx$$

The definite integral gives the area under the curve between $x = a$ and $x = b$.

Examples

1. Find the area of region bounded by the curve $y = 3x$, the x-axis and the vertical lines $x = 1$ and $x = 2$.

    Begin by sketching a graph.

    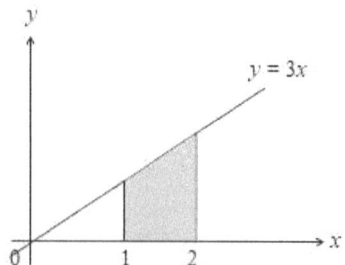

$$\text{Area} = \int_1^2 y\,dx \, dx$$
$$= \int_1^2 3x \, dx$$
$$= \left[\frac{3}{2}x^2\right]_1^2$$
$$= 6 - \frac{3}{2}$$
$$= \frac{9}{2}$$

The area is 9/2 square units.

You can use the formula of the triangle A = ½ bh to confirm this result. The area of the larger triangle minus the area of the smaller triangle gives the area A of the curve from x = 1 to x = 2.

$$A = \frac{1}{2} \cdot 2 \cdot 6 - \frac{1}{2} \cdot 1 \cdot 3 = \frac{9}{2}$$

2. Find the area enclosed by the curve $y = x^2 - 5x + 6$ and the x-axis.

   Sketch a graph.

   Find any x-intercept of the curve by solving $x^2 - 5x + 6 = 0$.

   $$x^2 - 5x + 6 = 0$$

   $$(x - 2)(x - 3) = 0$$

   giving $x = 2$ or $x = 3$.

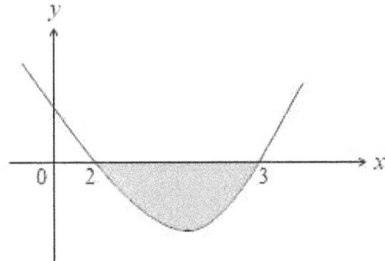

The area required is the area between $x = 2$ and $x = 3$.

$$\text{Area} = \int_2^3 y \, dx$$

$$= \int_2^3 (x^2 - 5x + 6) \, dx$$

$$= \left[\frac{1}{3}x^3 - \frac{5}{2}x^2 + 6x\right]_2^3$$

$$= \left(9 - \frac{45}{2} + 18\right) - \left(\frac{8}{3} - 10 + 12\right)$$

$$= \frac{9}{2} - \frac{14}{3}$$

$$= -\frac{1}{6}$$

Since area is nonnegative, the required area is given by 1/6 square unit.

3. Find the area enclosed by the curve $y = x^3 - 6x^2 + 8x$ and the x-axis.

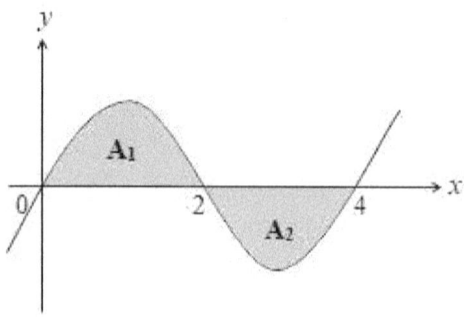

The area A_1 is above the x-axis whereas the area A_2 is below it. The area below the x-axis is negative. Find the area of the region above the x-axis and area below x-axis separately and then add the absolute values to get the total area.

$$A_1 = \int_0^2 (x^3 - 6x^2 + 8x)\, dx$$

$$= \left[\frac{1}{4}x^4 - 2x^3 + 4x^2\right]_0^2$$

$$= (4 - 16 + 16) - 0$$

$$= 4$$

$$A_2 = \int_2^4 (x^3 - 6x^2 + 8x)\, dx$$

$$= \left[\frac{1}{4}x^4 - 2x^3 + 4x^2\right]_2^4$$

$$= (64 - 128 + 64) - (4 - 16 + 16)$$

$$= 0 - 4$$

$$= -4$$

The total area = 4 + 4 = 8

The total area is 8 square units.

Using one integral over the entire interval to find this area would have given

$$A = \int_0^4 (x^3 - 6x^2 + 8x)\, dx$$

$$= \left[\tfrac{1}{4}x^4 - 2x^3 + 4x^2\right]_0^4$$

$$= (64 - 128 + 64) - 0$$

$$= 0$$

The values of two areas that are numerically equal and opposite had been added together in the process of integration, so they have cancelled each other out.

Exercise 15.3(a)

1. Find the areas enclosed by the x-axis, and the following curves and straight lines.

   (a) $y = 5x$, $x = 1$, $x = 4$

   (b) $y = x^2$, $x = 1$, $x = 3$

   (c) $y = 2x^2 - x + 1$, $x = -1$, $x = 2$

   (d) $y = x^2 + 3$, $x = -1$, $x = 2$

   (e) $y = (x-1)(x-3)$, $x = 0$, $x = 3$

   (f) $y = \frac{1}{x^2}$, $x = 1$, $x = 8$

2. Sketch the following curves and find the area enclosed by them, and by the x-axis, and the given straight lines.

   (a) $y = x(3-x)$, $x = 4$ \qquad (b) $y = -x^3$, $x = -2$

   (c) $y = x^2(x-1)$, $x = 2$ \qquad (d) $y = x^3 - 4x$, $x = 3$

3. Find the area below the curve $y = 6 - x - x^2$ and above the x-axis.

4. Find the area above the curve $y = x^2 - 4x + 3$ but below the x-axis.

5. Find the area below the curve $y = 1 - x^2$ and above the x-axis.

6. Sketch the curve $y = x^2 - 3x + 2$ and find the area cut off below the x-axis.

7. Sketch the curve $y = x(x + 2)(x - 2)$ and find the area of each of the two segments cut off by the x-axis.

8. Find the area below the curve $y = 2x^3 + 4x^2$, from $x = -2$ to $x = 0$.

9. Find the area between the y-axis, the curve $y = x^3 - 4$ and the line $y = 4$.

10. Find the area enclosed by the y-axis and the following curves and straight lines.

   (a) $y^2 = x, y = 3$ (b) $y^2 - x + 3 = 0, y = -1, y = 2$

   (c) $y = 4x^2, y = 1, y = 4$ (d) $y = x^3, y = 1, y = 27$

The Area between Two Curves

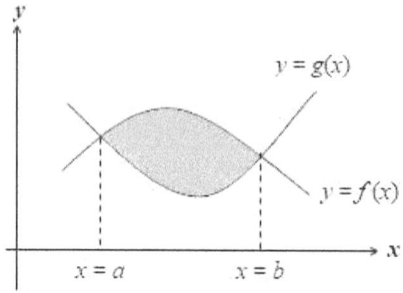

Figure 15.8

Figure 15.8 shows the graphs of $y = f(x)$ and $y = g(x)$. The area between the two curves between $x = a$ and $x = b$ (see shaded region) is the same as the area under the graph of $f(x)$ minus the area under the graph of $g(x)$. That is, the area between the graphs is given by

$$\int_a^b f(x)\, dx - \int_a^b g(x)\, dx$$

which can be written briefly as

$$\int_a^b [f(x) - g(x)]\, dx.$$

The limits of integration correspond to the x-coordinates of the points of intersection. To find the limits, we set the two function equal and solve for x.

Examples

Area under Curves **287**

1. Find the area enclosed by the curves $y = x^2$ and $y = 2x$.

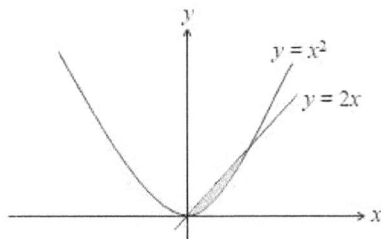

Begin by finding the limits of the integration.

$$x^2 = 2x$$

$$x(x - 2) = 0$$

giving $x = 0$ or $x = 2$.

The required area = $\int_0^2 (2x - x^2)\, dx$

$$= \left[x^2 - \frac{1}{3}x^3\right]_0^2$$

$$= \left(4 - \frac{8}{3}\right) - 0$$

$$= \frac{4}{3}$$

The area enclosed between the curves is 4/3 square units.

2. Find the area enclosed between the curves $y = x(x - 2)$ and $y = x(4 - x)$.

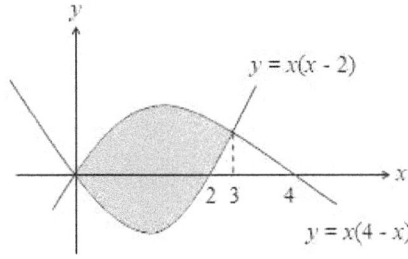

Begin by finding the limit of integration.

$$x(x - 2) = x(4 - x)$$

$$2x^2 - 6x = 0$$

$$2x(x - 3) = 0$$

giving $x = 0$ or $x = 3$.

The required area $= \int_0^3 [x(4 - x) - x(x - 2)]\, dx$

$$= \int_0^3 (6x - 2x^2)\, dx$$

$$= \left[3x^2 - \frac{2}{3}x^3\right]_0^3$$

$$= (27 - 18) - 0$$

$$= 9$$

The area enclosed between the curves is 9 square units.

Exercise 15.3(b)

1. Find the area of the segment cut off from each of the following curves by the given straight lines.

   (a) $y = x^2 - 3x - 4,\ y = 6$ (b) $y = x^2 + 2x - 3,\ y = 5$

   (c) $y = -x^2 + x + 6,\ y = -6$ (d) $y = x(x - 1),\ y = x$

   (e) $y^2 = x,\ y = x$ (f) $y = 8 - 2x - x^2,\ y + x - 2 = 0$

   (g) $y = x^2 - x - 6,\ y = 2x - 2$ (h) $y = x^2 + 1,\ y = 5$

2. Find the area enclosed by each of the following pairs of curves.

   (a) $y = 4 - x^2,\ y = x(x - 2)$

   (b) $y = x(x - 2),\ y = x(4 - x)$

   (c) $y = 2x^2 - 4x,\ y = x^2 - 3x$

   (d) $y^2 = x,\ y = x^3$

15.4 Kinematics

Kinematic is the study of motion. The position of an object relative to some fixed location at time t is called displacement and it is normally denoted by the letter s.

Velocity and Acceleration

A point P moves so that a small change in displacement Δs takes place over time Δt. Then, the average speed is given by $\Delta s/\Delta t$. The limit of $\Delta s/\Delta t$ as $\Delta t \to 0$ is ds/dt, called the instantaneous velocity. If v is the velocity of a particle at time t then

$$v = \frac{ds}{dt}.$$

The acceleration a, is the same as the derivative of velocity. Thus,

$$a = \frac{dv}{dt} = \frac{d^2 s}{dt^2}$$

Unless otherwise stated, distance and time are measured in metres (m) and second (s) respectively. The velocity of an object is measured in metres per second (m s^{-1}) and acceleration in metres per second squared (m s^{-2}).

Examples

1. A particle moves along a straight line so that its distance in metres after t s is $s = 3t^2 + 2t$. Find its distance, velocity and acceleration after 5 seconds.

$$s = 3t^2 + 2t$$

When $t = 5$

$$s = 3 \cdot 5^2 + 2 \cdot 5$$
$$= 75 + 10$$
$$= 85$$

The distance moved is 85 m.

The velocity of the particle is given by

$$\frac{ds}{dt} = 6t + 2$$

When $t = 5$

$$\frac{ds}{dt} = 6 \cdot 5 + 2$$

$$= 32$$

The velocity is 32 m s^{-1}.

The acceleration is given by

$$\frac{d^2s}{dt^2} = 6$$

The acceleration is 6 m s^{-2}.

2. A ball is thrown vertically upwards and its height after t s is s m, where $s = 29.4t - 4.9t^2$. Find when the ball is momentary at rest and the greatest height reached.

$$s = 29.4t - 4.9t^2$$

$$\frac{ds}{dt} = 29.4 - 9.8t$$

The ball would be at rest when $ds/dt = 0$

$$0 = 29.4 - 9.8t$$

giving $\qquad t = 3$

The ball would be at rest at 3 s.

The ball reaches the greatest height at $t = 3$, so

$$s = 29.4 \cdot 3 - 4.9 \cdot 3^2$$

$$= 44.1$$

The greatest height reached is 44.1 m.

Exercise 15.4(a)

1. A particle moves along a straight line so that after t s, its distance from a fixed point on the line is s m, where $s = 2t^3 - 3t^2 + 4$. Find the distance, velocity and acceleration after 3 s.

2. A body moves so that its distance from the starting point in metres is $s = t^3 + 3t$ after t seconds. Find the distance, velocity and acceleration after 2 s.

3. The distance s m of a moving point A at time t s is given by $s = 6t^2 - t^3$. Find the total distance moved during the 2 nd second and the maximum distance moved.

4. The distance s m of a moving particle at time t s is given by $s = t^4 - 3t^2 + 4t$. Find the velocity and acceleration of the particle at the instant $t = 2$ s.

5. A ball is thrown vertically upwards and its height after t s is s m, where $s = 19.6t - 4.9t^2$. Find when the stone is momentarily at rest. What is the greatest height reached?

6. A particle moves from O towards A. It is s m from O after t s, where $s = t(t - 3)$. When is it again at O? What is the particle's greatest displacement from O?

7. A particle moves along a straight line so that its distance from O a fixed point on the line is s m, where $s = t^3 - 5t^2 + 6t$. When is the particle at O? What is its velocity and acceleration at these times?

8. A particle P is travelling along a straight line so that its distance from the starting point s m after time t s is given by the equation $s = 2t^2 - \frac{1}{3} t^3$. Calculate the velocity and acceleration of P after 3 s, and the distance travelled by P when it first comes to rest.

9. A particle is moving in a straight line and its distance s m from a fixed point in the line after t s is given by $s = 18t - 21t^2 + 4t^3$. Find:

    (a) the velocity and acceleration of the particle after 3 s.

    (b) the distance travelled between the two times when the velocity is instantaneously zero.

10. A bus which runs from P to Q stops at two adjacent rest stops. The distance s km, travelled by the bus t hours after passing a railway bridge between the two rest stops is $s = 9t + 3t^2 - t^3$. Find the distance of each rest stop from the railway bridge.

11. A bus which stops at two adjacent bus stops passes a police check point at 8 a.m. After t hours the bus is s km from the check point where $s = 6t + 3t^2 - 4t^3$. Find the time of departure from the stop and the time of arrival at the second point.

Suppose a particle has an acceleration a at time t, then its velocity is the integral of acceleration, and its displacement is the integral of velocity. That is,

$$v = \int a\, dt \quad \text{and} \quad s = \int v\, dt$$

Examples

1. A particle starting from rest at O moves along a straight line OA so that its acceleration after t s is $(3t^2 - 4t)$ m s^{-2}. Find its displacement after 3 seconds.

Let a be the acceleration of the particle at time t.

$$a = 3t^2 - 4t$$

$$v = \int (3t^2 - 4t)\, dt$$

$$= t^3 - 2t^2 + c$$

Since the particle starts from rest $v = 0$ when $t = 0$. So,

$$c = 0$$

Hence

$$v = t^3 - 2t^2$$

The distance s is given by

$$s = \int (t^3 - 2t^2)\, dt$$

$$= \frac{1}{4}t^4 - \frac{2}{3}t^3 + c$$

When $t = 0$, $s = 0$, $c = 0$.

Hence

$$s = \frac{1}{4}t^4 - \frac{2}{3}t^3$$

When $t = 3$

$$s = \frac{1}{4} \cdot 3^4 - \frac{2}{3} \cdot 3^3$$

$$= \frac{9}{4}$$

$$= 2.25$$

The displacement of the particle is 2.25 m.

2. A stone is thrown vertically upwards from the ground level with a velocity of 24.5 m s^{-1}. If the acceleration due to gravity is 9.8 m s^{-1}, find the highest height reached.

Let a be the acceleration of the particle.

$$a = -9.8$$

Thus
$$v = \int -9.8 \, dt$$

$$= -9.8t + c$$

When $t = 0$, $v = 24.5$ and $c = 24.5$

Hence
$$v = -9.8t + 24.5$$

At the highest height $v = 0$

$$0 = -9.8t + 24.5$$

giving $t = 2.5$

Now,
$$s = \int (-9.8t + 24.5) \, dt$$

$$= -4.9t^2 + 24.5t + c$$

When $t = 0$, $s = 0$ and $c = 0$

Hence
$$s = -4.9t^2 + 24.5t$$

When $t = 2.5$, we have

$$s = -4.9 \cdot 2.5^2 + 24.5 \cdot 2.5$$

$$= 30.6$$

The highest height reached is 30.6 m.

Exercise 15.4(b)

1. Find the displacement s m of a particle at time t s if its velocity is $(3t^2 + 4t)$ m s^{-1}, and $s = 2$ and $t = 0$.

2. Find the velocity v m s^{-1} and displacement s m of a particle at time t s, if its acceleration a is $(6t - 8)$ m s^{-2}, and $s = 4$, $v = 6$ when $t = 0$.

3. A particle moves with a velocity equal to $(2t^2 - 3t)$ m s^{-1}, where t is the time in seconds. If it starts at the point P, find the distance from P after 6 seconds.

4. A particle has an acceleration $(18t + 3)$ m s^{-2} and starts from rest. What is its velocity after 2 s and how far has it moved?

5. A particle has an acceleration of $6t$ m s^{-2}, where t is the time in seconds. It starts with velocity of 2 m s^{-1} from O. What is its velocity and distance from O after 3 seconds?

6. If the velocity of a particle is $v = 64 - t^3$ in m s^{-1}, and it starts so that $s = 0$ when $t = 0$. Find the distance from the starting point when the velocity is zero for an instant.

7. The velocity of a particle is $(3t^2 + 2t)$ m s^{-1}, and it starts + 6 m from the point O when $t = 0$. What is its acceleration and distance from the point O after 5 s?

8. A particle moves so that its velocity is given by $(12t - t^2)$ m s^{-1}, where t is the time in seconds. If $s = 0$ when $t = 0$, find the distance from the starting point where acceleration is zero.

9. A particle starts from rest at O and moves along a straight line. After t s its acceleration is a m s^{-2}, where $a = 3t^2 - t^3$. When was the particle momentary at rest?

10. A bus starts at a terminal and stops at a rest stop t hours after it leaves the terminal, with a speed of $(240t - 150t^2)$ km h^{-1}. Find the distance between the terminal and the rest stop.

11. A ball is thrown vertically downwards at 20 m s^{-1} from the fifth floor of a building. If the acceleration due to gravity is 9.8 m s^{-2}, find its velocity and position after 2 s.

12. A ball is thrown vertically upwards from the ground level with a velocity of 25 m s^{-1}. If the acceleration due to gravity is 10 m s^{-2}, find the greatest height reached.

15.5 Distance and Velocity Time Graphs

The velocity-time graph relates the velocity to the time.

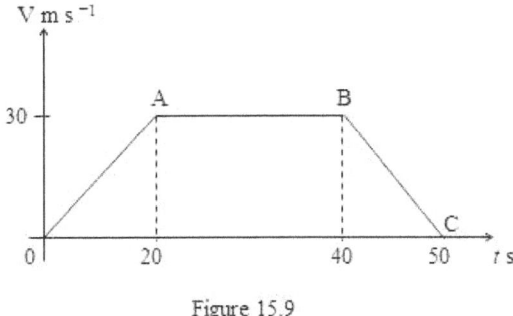

Figure 15.9

Figure 15.9 shows the velocity-time graph of a particle. The graph shows that the particle starts from rest, moving with a steady increasing velocity to 30 m s^{-1}. Then moves with a constant velocity between 20 and 40 seconds, and finally has a steadily decreasing speed between 40 and 50 seconds.

The gradient of a velocity-time graph is the acceleration of the particle. You can see that the acceleration is uniform between 0 and 20 seconds and between 40 and 50 seconds. The gradient of OA gives the acceleration as 30/20, i.e.1.5 m s^{-2}. The line BC slopes downwards with a negative gradient. This represents an object that is steadily slowing down. The acceleration is negative. A negative acceleration is called a deceleration (or retardation). In this case deceleration is 30/10, i.e.3 m s^{-2}.

The area under velocity-time graph is the displacement. The graph and t- axis formed a trapezium. The area of the trapezium gives the total distance travelled by the particle, and the height of the trapezium is the maximum velocity.

Note that the distance moved in each phase of the journey is equal to the area between the corresponding graph and the time-axis. The area of the trapezium is given by A = h(a + b)/2, where a and b are the length of the parallel sides and h is the height of the trapezium. In this case, $a = 50$, $b = 20$ and $h = 30$. Substituting these values gives

$$\text{Distance} = \tfrac{1}{2} \cdot 30 \cdot (50 + 20)$$

$$= 1050 \text{ m}$$

Alternatively, we can break the trapezium into two triangles and a rectangle. The area of the trapezium is the same as the sum of the area of each triangle and the area of the rectangle. In this case, we have

$$\text{Distance} = \tfrac{1}{2} \cdot 30 \cdot 20 + 20 \cdot 30 + \tfrac{1}{2} \cdot 10 \cdot 30$$

$$= 300 + 600 + 150$$

$$= 1050 \text{ m}$$

Examples

1. A car travels on a horizontal road from village A to another village B. The car starts from rest and accelerates uniformly for 20 second reaching a speed of 16 ms^{-1}. It maintains this speed for 15 seconds and then decelerates uniformly coming to rest in 15 seconds. Find

   (a) the distance AB

   (b) the average speed of the journey

   (c) the acceleration during the first 20 seconds

   The motion of the car is shown in the velocity-time graph.

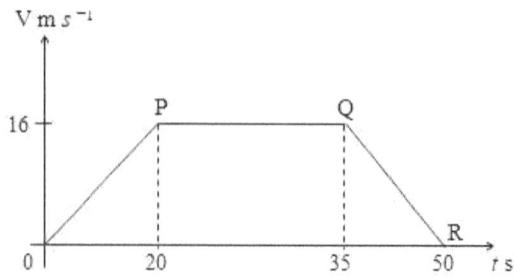

(a) Area of the trapezium $= \tfrac{1}{2} \cdot 16 \cdot (15 + 50)$

$$= 520$$

(b) Average speed = $\dfrac{total\ distance}{time}$

$= \dfrac{520}{50}$

$= 10.4$

The average speed is 10.4 ms^{-1}

(c) The acceleration is given by the gradient of the line OP.

$$\text{Gradient} = \dfrac{16}{20}$$

$= 0.8$

The acceleration during the first 20 seconds is 0.8 ms^{-2}.

2. A train at a station P accelerates uniformly from rest until it attains a speed of 150 km h^{-1}. It then continues at this speed for some time and decelerates uniformly until it comes to a stop at a station Q 120 km from P. The total time taken for the journey is 1 hour. If the rate of deceleration is twice that of acceleration calculate:

(a) the time taken during which the constant speed is maintained.

(b) the acceleration of the train.

(a)

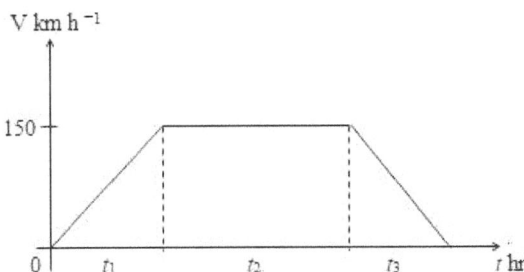

Suppose the train takes t_1, t_2 and t_3 hours for the three parts of the journey as indicated in the diagram above. The train travels at constant speed for t_2 hours. The area under the graph represents the distance travelled.

Therefore $\frac{1}{2} \cdot (t_2 + 1) \cdot 150 = 120$

$$t_2 + 1 = \frac{8}{5}$$

$$t_2 = \frac{3}{5}$$

Hence the train travels at constant speed for 3/5 hours or 36 minutes.

(b) The time taken to decelerates is ½ the time it takes to accelerates so $t_3 = t_1/2$. We know that the journey takes 1 hour so

$$t_1 + \frac{3}{5} + \frac{1}{2}t_1 = 1$$

$$\frac{3}{2}t_1 = \frac{2}{5}$$

$$t_1 = \frac{4}{15}$$

Hence the time it takes to accelerates is 4/15 hours or 16 minutes. Therefore,

$$a = \frac{150}{\frac{4}{15}}$$

$$= 562.5$$

Hence the acceleration is 562.5 km h⁻².

Alternatively, using the equation $v = u + at$ the acceleration a m s⁻² is given by

$$150 = 0 + a \cdot \frac{4}{15}$$

giving $\qquad a = 562.5$ km h⁻².

Exercise 15.5

1. A vehicle accelerates uniformly for 6 seconds, travels at a constant speed for 15 second, and then decelerates uniformly to rest in 4 seconds. If the total distance travelled is 800 metres, find the maximum speed.

2. A car accelerates at 3 ms^{-2} for 5 seconds, then travels at a constant speed for 16 seconds, and then decelerates uniformly to rest. The total distance travelled is 300 m. Draw a velocity-time graph and hence find the deceleration.

3. A cyclist starts from rest, and accelerates uniformly to a maximum speed of 15 ms^{-1}. This speed is maintained for 20 seconds, and then the cyclist decelerates uniformly to rest in 2 s. The total distance covered is 375 metres. Sketch a velocity-time graph. Find the total time taken and the acceleration.

4. A train starts from rest at a station P and accelerates uniformly until it reaches a velocity of 36 km h^{-1} after 3 hours. It maintains this velocity for some time and then retards uniformly for 2 hours to come to rest at station Q. If the distance between P and Q is 270 kilometres, find:

   (a) the time the train travelled a maximum velocity

   (b) the acceleration of the train

   (c) the retardation of the train

5. A train starts from rest from station P and accelerates uniformly for 3 minutes reaching a speed of 60 km h^{-1}. It maintains this speed for 5 minutes and then retards uniformly for 2 minutes to come to rest at station Q. Find:

   (a) the distance PQ in kilometres

   (b) the average speed of the train

   (c) the acceleration in ms^{-2}

6. A train starts from station X and accelerates uniformly to a speed of 20 m s^{-1}. It maintains this speed and then retards uniformly until it comes to rest at a station Y. The distance between the stations is 2 kilometres. The total time taken is 2 minutes. If the retardation is twice the acceleration in magnitude, find:

   (a) the time for which the train is travelling at constant speed

   (b) the acceleration of the train

Review Exercise 15

1. Differentiate the following with respect to x.

   (a) $y = 3x^2 + 2x$

   (b) $y = 2 + 2x - 4x^2$

   (c) $y = x^4 + x$

   (d) $y = x^4 - x^2 + 2$

   (e) $y = 1 + \frac{2}{\sqrt{x}} - \frac{3}{x}$

   (f) $y = 2x^5 - \frac{4}{x^4} + \frac{3}{\sqrt[3]{x}}$

2. Differentiate the following with respect to x.

   (a) $y = x(x - 2)$

   (b) $y = (3 - 2x^2)(2 + x^3)$

   (c) $y = (3 - 2x^2)^2$

   (d) $y = \frac{3x^2 + 2x + 1}{x}$

   (e) $y = \frac{x-8}{\sqrt{x}}$

   (f) $y = \frac{\sqrt[3]{x} + \sqrt[4]{x} - 3}{\sqrt{x}}$

3. Differentiate the following with respect to x.

   (a) $y = (3x - 7)^{1/2}$

   (b) $y = (2 + x^2)^5$

   (c) $y = \sqrt{1 - 2x^2}$

   (d) $y = x(1 - 2x)^{1/2}$

   (e) $y = \frac{x-4}{\sqrt{x^2-1}}$

   (f) $y = \frac{\sqrt{x}}{\sqrt{x-1}}$

4. Differentiate the following with respect to x.

   (a) $y = \sin 2x$

   (b) $y = \cos 5x$

   (c) $y = 2\sin 3x$

   (d) $y = -\frac{1}{2}\cos 6x$

   (e) $y = \sin(3x - 2)$

   (f) $y = 6\sin\frac{1}{2}x$

   (g) $y = \cos(2x + 3)$

   (h) $y = \tan(3x + 5)$

   (i) $y = \cos x^2$

   (j) $y = \tan(x^2 - 1)$

   (k) $y = \sin(x^2 + 3)$

   (l) $y = \tan 3x^2$

5. Differentiate the following with respect to x.

(a) $y = x \sin x$

(b) $y = x \cos 2x$

(c) $y = 2x \tan x$

(d) $y = x^2 \sin x$

(e) $y = x^2 \tan 3x$

(f) $y = x^3 \cos x$

(g) $y = \dfrac{\cos x}{x}$

(h) $y = \dfrac{\sin 2x}{x}$

(i) $y = \sin 2x \cos x$

(j) $y = \cos x \tan 3x$

6. Differentiate the following with respect to x.

(a) $y = \ln \dfrac{1}{4} x$

(b) $y = \ln 2x$

(c) $y = 3 \ln \dfrac{1}{2} x$

(d) $y = \ln(2x - 3)$

(e) $y = \ln\left(\dfrac{1}{x^2}\right)$

(f) $y = \ln\left(\dfrac{3}{x^3}\right)$

7. Differentiate the following with respect to x.

(a) $y = e^{-3x}$

(b) $y = e^{\frac{1}{3}x}$

(c) $y = 4e^{\frac{1}{4}x}$

(d) $y = e^{4x^3}$

(e) $y = e^{2x+3}$

(f) $y = e^{\sqrt{x}-4}$

8. Differentiate the following with respect to x.

(a) $y = x \ln x$

(b) $y = x^2 \ln x^3$

(c) $y = 2x^3 \ln(2x - 1)$

(d) $y = xe^{3x}$

(e) $y = x^2 e^{\frac{1}{2}x^2}$

(f) $y = x \ln \dfrac{1}{2}(x^2 + 1)$

9. For each of the following functions express dy/dx in terms of x and y.

(a) $x^2 - y^2 = 4$

(b) $x^2 + y^2 + 3x - y = 0$

(c) $x^2 - 4xy + y^2 = 5$

(d) $x^2 - 3x^3 y - 4y = 0$

10. Find the second derivative of the following.

   (a) $y = 2x^2 + 3x - 1$
   (b) $y = 6x^2 - 3x + 5$

   (c) $y = -2x^3 + 4x + 7$
   (d) $y = (x^3 - 1)(x - 4)$

   (e) $y = 3\sqrt{x} + 2x^2$
   (f) $y = (x + 1)^3$

11. Find the differential coefficient of $y = 2x^3 + 3x^2 - 4x - 1$ and determine the gradient of the curve at $x = 2$.

12. The gradient of a curve is given by $2x - 3x^2$. Find the equation of the curve if $(1, 2)$ lies on it.

13. Find the equations of the tangents and normals to the following curves at the points given.

   (a) $y = 2x^2 - 3x + 1$, $(-1, 6)$
   (b) $y = x^3 + 2x$, $(1, 3)$

   (c) $y = x^3 + 2x^2$, $(-2, 0)$
   (d) $y = 1 + \frac{2}{x}$, $(2, 2)$

14. Find the equation of the tangent of the curve $y = 6 - x - x^2$ at the point where $x = 2$.

15. Find the equation of the tangent to the curve $y = x^3 - 11x$ at the point $(1, -10)$ and also the equation of the tangents which are parallel to the line $x - y = 0$.

16. Find the equation of the tangent at the point $(2, 2)$ on the curve $y = x^3 - 3x$ and the coordinates of the point at which this tangent meets the curve again.

17. Find the gradients of the following functions at the points shown.

   (a) $x^2 - y^2 = 5$, $(2, 1)$
   (b) $x^2 + y^3 = 1$, $(-3, 2)$

   (c) $x^2 + y^2 - 3x + 2y = 1$, $(2, 1)$
   (d) $x^2 - 2xy = 5$, $(-1, 2)$

   (e) $x^2 y + x^3 + 2y = 2$, $(2, -1)$
   (f) $x^2 y^2 + 2x - 3y = 0$, $(1, 1)$

18. Find any maximum or minimum points on the following curves.

   (a) $y = x^2 + 2x - 2$
   (b) $y = 3 - 2x - x^2$

(c) $y = x^3 - 3x$ (d) $y = x(x^2 - 12)$

(e) $y = x^3 - 2x^2 - 4x$ (f) $y = x^3 + x^2 - x$

19. Sketch the curve of the following functions.

(a) $y = x^2 - 2x$ (b) $y = 3 - 2x - x^2$

(c) $y = 3x - x^3$ (d) $y = x^2(x - 3)$

(e) $y = x^2(3 - 2x)$ (f) $y = x^2(x^2 - 8)$

20. A metal basket is made in the form of a cylinder with an open top. Its height is h cm and its radius is r cm. It is to contain 2000 cm^3 of liquid. What is the radius, if it is to consist of the smallest possible amount of metal?

21. A cardboard box is to be made in the form of a cuboid with square cross-section. Its volume must be 8000 cm^3. Let x cm be the side of the square and x cm be the length. What value of x will use the least cardboard?

22. 100 cm of wire is cut into two pieces, one of which is made into a square and the other into a circle. Find the radius of the circle if the sum of the two areas is to be as small as possible.

23. A three-sided fence is to be built by a farmer next to a straight section of a river which forms the fourth side of a rectangular field. The side of the fence parallel to the river will cost £ 5 per metre to build, whereas the sides perpendicular to the river will cost £ 3 per metre. If the enclosed area is equal to 450 m^2, find the minimum perimeter and the cost of fencing the field.

24. A cube has side 3 cm. Find the approximate change in its volume if its side changes by 0.1 cm.

25. The side of a square is measured with a possible error of 2%. What is the approximate error in areas?

26. The percentage error when measuring the area of a circle was 5%. What was the approximate percentage error in its radius?

27. The side of a square is increasing at 0.5 cm s^{-1}. At what rate is the area increasing at a time when the side is 25 cm?

28. The radius of a sphere is increasing at 0.02 cm s^{-1}. At what rate is the surface area increasing when the radius is 8 cm?

29. Water is poured into a cone, of semi-vertical angle of 45°, at what rate is it increasing?

30. For each of the following find y as a function of x.

   (a) $\frac{dy}{dx} = 2x$, $y = 1$ for $x = 0$

   (b) $\frac{dy}{dx} = 3x^2 + x$, $y = 1$ for $x = 1$

   (c) $\frac{dy}{dx} = x^2 + x + 1$, $y = 3$ for $x = 1$

   (d) $\frac{dy}{dx} = x(x^2 + 1)$, $y = 0$ for $x = 1$

31. Find the following integrals.

   (a) $\int 2x \, dx$ (b) $\int (3x^2 + 2x + 1) \, dx$

   (c) $\int (5 + x^2 + 2x^3) \, dx$ (d) $\int x(3x^2 + 7) \, dx$

   (e) $\int (x - 3)(x^2 + 2x) \, dx$ (f) $\int x(4x^2 - 2x^3) \, dx$

32. Find the following integrals.

   (a) $\int \cos 3x \, dx$ (b) $\int \sin 2x \, dx$

   (c) $\int 4 \sec^2 2x \, dx$ (d) $\int \sin(2x + 1) \, dx$

   (e) $\int \sec^2(4x - 3) \, dx$ (f) $\int 4 \cos\left(\frac{1}{2}x + 3\right) dx$

33. Find the following integrals.

   (a) $\int 3e^{2x} \, dx$ (b) $\int \frac{5}{2} e^{-2x} \, dx$ (c) $\int 4e^{3x} \, dx$

   (d) $\int \frac{9}{2} e^{-3x+4} \, dx$ (e) $\int 3e^{3x-4} \, dx$ (f) $\int e^{\frac{1}{2}x+1} \, dx$

34. Evaluate the following integrals.

   (a) $\int_0^3 (2x + 3x^2) \, dx$ (b) $\int_1^2 \left(x^2 - \frac{1}{6}x\right) dx$

(c) $\int_1^9 \left(\sqrt{x} - \frac{3}{x^2}\right) dx$ (d) $\int_1^4 x(x+1)\, dx$

(e) $\int_{-1}^2 (x^3 + 2)^2\, dx$ (f) $\int_1^4 \frac{x^2 - 1}{\sqrt{x}}\, dx$

35. Find the displacement s m of a particle at time t s if its velocity is $(3t^2 + 4t)$ m s^{-1} and $s = 2$ when $t = 0$.

36. Find the velocity v m s^{-1} and displacement s m of a particle at time t s if its acceleration a is $(6t - 8)$ m s^{-2} and $s = 4$, $v = 6$ when $t = 0$.

37. Find the velocity v m s^{-1} and displacement s m of a particle at time t s if we know that its acceleration a m s^{-2} is given by:

(a) $a = 3t - 4$, and when $t = 0$, $s = 5$ and $v = 6$

(b) $a = -10$, and when $t = 1$, $s = 6$ and $v = 3$.

38. A particle starts from rest at O and moves along a straight line. After t s its velocity is v m s^{-1}, where $v = 2t - t^2$. Show that the particle is momentarily at rest after 2 s and find its distance from O at this time. Find also the maximum velocity of the particle during the first 2 s of the motion.

39. A particle starts from rest at a point 6 m from O and moves in a straight line away from O with a velocity v m s^{-1} at time t s given by $v = t - (1/12)t^2$. Find:

(a) its acceleration and distance from O, each in terms of t.

(b) the time at which it begins to return, and the time at which it again reaches its starting point.

40. Find the areas enclosed by the following curves and the x-axis between the limits shown.

(a) $y = x$, $x = 1$ and $x = 3$

(b) $y = \frac{1}{x^2}$, $x = 1$ and $x = 20$

(c) $y = x^{-1/2}$, $x = 0$ and $x = 9$

(d) $y = x^3 - x$, $x = 0$ and $x = 1$

41. A particle accelerates uniformly for 2 seconds, travels at a constant speed for 12 seconds, then decelerates uniformly to rest in 4 seconds. If the total distance travelled is 750 m, find the maximum speed.

42. A car accelerates at 3 m s^{-2} for 4 seconds, then travels at a constant speed for 20 seconds, and then decelerates uniformly to rest. The total distance travelled is 300 m. Draw a velocity-time graph and hence find the deceleration.

43. A cyclist starts from rest and accelerates uniformly to a maximum speed of 15 m s^{-1}. This speed is maintained for 30 seconds, and then the cyclist decelerates uniformly to rest. The total distance covered is 525 m. Sketch a velocity-time graph. Find the time taken.

44. A train accelerates uniformly from rest at a station P until it attains a speed of 100 km h^{-1}. This speed is maintained for some time and then the train decelerates uniformly until it comes to a stop at station Q. The distance between the stations is 60 km and the time taken for the journey is 1 hour. If the rate of deceleration is twice that of acceleration, calculate the time for which the speed is constant and the acceleration of the train.

Answers to Exercises

Chapter 1

Exercise 1.1

1. {5, 6, 7, 8, 9, 10, 11, 12}   2. {12}   3. {17, 19, 23, 29}

4. {−4, −3, −2, −1, 0, 1, 2}   5. {9}   6. True   7. False

8. True   9. True   10. True   11. False   12. True

13. Infinite   14. Finite   15. Infinite   16. Finite   17. Infinite

18. Finite   19. 4   20. 1   21. 12   22. 5

23. Not empty   24. Empty   25. Not empty   26. Empty

27. ⊂   28. ⊄   29. ⊂   30. ⊂   31. ⊄   32. ⊆

33. ⊄   34. ∅, {2}   35. ∅, {0}, {1}, {0, 1}

36. ∅, {−2}, {1}, {3}, {−2, 1}, {−2, 3}, {1, 3}, {−2, 1, 3}

37. 16   38. 32   39. 128   40. 128

Exercise 1.2

1. (a) {1, 2, 3, 4, 5, 6, 7, 9, 10}   (b) {1, 2, 3, 4, 5, 6, 7, 9, 10}

2. (a) {−4, −3, −2, −1, 0, 1, 2, 3, 4, 5}   (b) {−4, −3, −2, −1, 0, 1, 2, 3, 4, 5}

3. (a) {0, 1, 2, 5, 7, 10}   (b) {0, 1, 2, 5, 7, 10}   4 (a) {−1, 0, 1, 7}   (b) {−1, 0, 1, 7}

5. (a) {2, 3, 5, 7}   (b) {2, 3, 5, 7}   6. (a) {−3, −2, −1, 0, 2, 3, 4, 5, 6}

(b) {−3, −2, −1, 0, 2, 3, 4, 5, 6}   7. (a) {3}   (b) {3}

8 (a) {−2, 0, 1, 2, 3, 4, 5}   (b) {−2, 0, 1, 2, 3, 4, 5}

9. (a) {−1, 1, 2, 5}   (b) {−1, 1, 2, 5}   10 (a) {2, 5, 8, 10}   (b) {1, 3, 6, 8, 9}

11. (a) {2, 4, 6, 7, 8, 9, 10}   (b) {1, 3, 6, 8, 9}   12(a) {d}   (b) {d}

13. (a) {2, 4, 5, 7, 8, 9}   (b) {2, 4, 5, 7, 8, 9}   14. B ⊂ A   15. $P \cap R$

16. $A \cap B = \emptyset$ 17. $H \cap P = \emptyset$ 18. $G \cap R$

Exercise 1.3

1. (a) (b) (c)

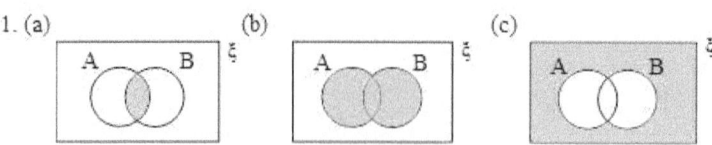

2. (a) (b) (c)

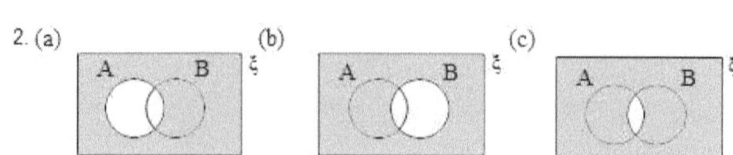

3. (a) (b)

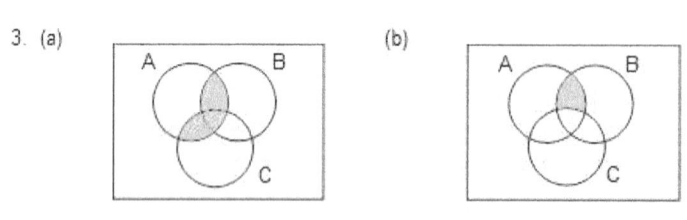

4. (a) (b)

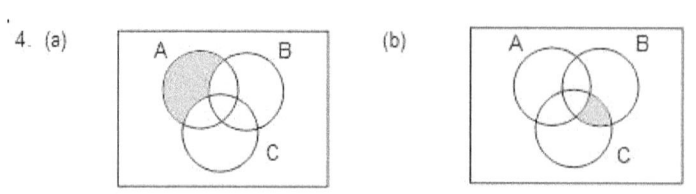

Review Exercise 1

1. $\{6, 7, 8, 9, 10, 11, 12\}$ 2. $\{9, 12\}$ 3. $\{2\}$ 4. $\{1, 2, 3, 4, 6, 9, 18, 36\}$

5. $\{1, 4, 9\}$ 6. True   7. False   8. False   9. True   10. False   11. True

12. Finite   13. Infinite   14. Finite   15. Infinite   16. 8   17. 5   18. 3

19. 5   20. Not empty   21. Empty   22. Not empty   23. Not empty   24. $\in$

25. $\notin$ 26. $\nsubseteq$ 27. $\subset$ 28. $\subset$ 29. $\in$ 30. $\nsubseteq$ 31. $\subset$

32. (a) $\{3, 4, 5, 6, 7, 8\}$ (b) $\{1, 2, 3, 4, 5, 7, 8\}$ (c) $\{1, 2, 3, 4, 5, 6, 7, 8\}$

33. (a) $\{-1, 0, 3\}$ (b) $\{-2, -1, 3\}$ (c) $\{-1, 3\}$

34. (a) {1, 7, 8, 9, 10} (b) {1, 2, 3, 4, 9} c) {1, 9} (d) {1, 2, 3, 4, 7, 8, 9, 10}

35. (a) 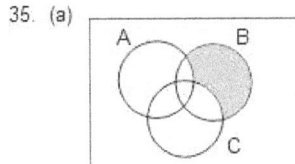 (b)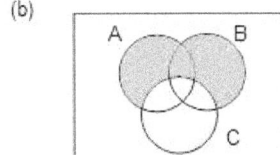

(c)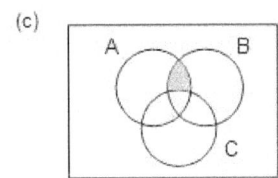

Chapter 2

Exercise 2.1

1. (a) Not a function (b) Function (c) Not a function (d) Function

2. (a) Function (b) Not a function (c) Function (d) Not a function

3. (a) Domain: all real numbers

   Range: all real numbers greater than or equal to 0.

   (b) Domain: all real numbers greater than or equal to 0

   Range: all real numbers greater than or equal to 0

   (c) Domain: all real numbers

   Range: all real numbers less than or equal to 2

   (d) Domain: all real numbers   Range: all real numbers

4. (a) Domain: all real numbers greater than or equal to 0

   Range: all real numbers greater than or equal to −2

   (b) Domain: all real numbers   Range: all real numbers

   (c) Domain: all real numbers   Range: all real numbers

   (d) Domain: all real numbers

Range: all real numbers greater than or equal to 3

5. (a) 10 (b) 5 (c) 13 (d) 26   6. (a) −5 (b) 0   c) 4 (d) 3

7. (a) 7/8 (b) −5/4 (c) 20 (d) 26/3

8. (a) 5 (b) 1 (c) 5 (d) 19   9. −2

10. −5 or 3   11. −23/7   12. 11

Exercise 2.2

1. (a) $3x^2 + 5$ (b) $9x^2 + 12x + 3$

2. (a) $7 − 6x$ (b) $4x^2 + 3$ (c) $2 − 6x^2$ (d) $−6x − 7$

3. (a) $10 − 6x^2 + x^4$ (b) $11 − 2x^2$ (c) $4x^2 + 20x + 26$
   (d) $−6 + 6x − x^2$ (e) $x^4 + 2x^2 + 2$

4. (a) $9x^2 − 24x + 16$ (b) $3x^2 − 4$ (c) $9x^2 + 12x + 2$ (d) $3x^2$

5. (a) 26 (b) 196 (c) 11   6. (a) $−19/2$ (b) 3

7. (a) 9 (b) 48 (c) 5 (d) 63   8. (a) 2 (b) 3

9. (a) −2, 1 −5, 5   10. −1, 2/3

Exercise 2.3

1. (a) Not one-to-one (b) One-to-one (c) One-to-one
   (d) Not one-to-one (e) One-to-one (f) One-to-one

2. (a) One-to-one (b) One-to-one (c) One-to-one (d) One-to-one

3. (a) Not one-to-one (b) One-to-one (c) One-to-one
   (d) One-to-one (e) One-to-one (f) Not one-to-one

Exercise 2.4

1. (a) Has inverse (b) Has no inverse (c) Has inverse
   (d) Has inverse (e) Has no inverse (f) Has inverse

2. (a) Yes (b) Yes (c) No (d) No (e) Yes

3. (a) $f^{-1}(x) = \frac{1}{3}(x + 8)$ (b) $g^{-1}(x) = \sqrt{4 - x}, \; x \le 4$

(c) $h^{-1}(x) = \frac{5-x}{2x}, \; x \ne 0$ (d) $f^{-1}(x) = \frac{4x+3}{x-2}, \; x \ne 2$

(e) $g^{-1}(x) = x^3 - 2$ (f) $h^{-1}(x) = \sqrt{x} - 3, \; x \ge 0$

Exercise 2.5

1. −8, 4   2. −12, 2   3. 7/2, 23/2   4. −2, 24/7

5. No solution   6. −4, 72/7   7. No solution   8. 1/3, 1

9. 2/3, 2   10. −3, 7   11. 1, 5/3   12. −3, 1   13. 1/2

14. 1   15. −17/2, −3/4   16. −3, 1   17. 11, 13

18. −11/4, 3/8   19. −29/8, 21/2   20. −5/6

Review exercise 2

1. (a) Not a function   (b) Function   (c) Not a function

(d) Function   (e) Function   (f) Function

2. (a) Domain: all real numbers

　　Range: all real numbers greater than or equal to −3.

　(b) Domain: all real numbers greater than 0.

　　Range: all real numbers less than 4.

(c) Domain: all real numbers greater than −4 and less than or equal to 3

　Range: all real numbers greater than −3 and less than or equal to 2.

(d) Domain: all real numbers.

　Range: all real numbers less than or equal to 3.

(e) Domain: all real numbers greater than 0.   Range: all real numbers

(f) Domain: all real numbers   Range: all real numbers

3. (a) Domain: all real numbers   Range: all real numbers

　(b) Domain: all real numbers.

Range: all real numbers greater than or equal to -2.

(c) Domain: all real numbers less than or equal to 3.

Range: all real numbers greater than or equal to 0.

(d) Domain: all real numbers.

Range: all real numbers greater than or equal to 2.

4. (a) 2 (b) -7 (c) 7   5. (a) 24 (b) $33/4$ (c) -8

6. (a) $6x^2 + 1$ (b) $12x^2 - 36x + 29$ (c) $4x - 9$ (d) $27x^4 + 36x^2 + 14$

7. (a) $-2 - x^2 + x^4$ (b) $3x^2 - 9x + 2$ (c) $-9x^2 - 12x - 2$
(d) $-x^4 + 6x^3 - 9x^2 + 2$ (e) $9x^2 + 3x - 2$ (f) $8 - 3x^2$
(g) $x^4 - 6x^3 + 6x^2 + 9x$ (h) $-2 + 4x^2 - x^4$ (i) $9x + 8$

8. (a) -43 (b) 75 (c) 27 (d) -103

9. (a) One-to-one (b) Not one-to-one (c) One-to-one (d) Not one-to-one
(e) One-to-one (f) One-to-one

10. (a) Yes (b) Yes (c) No (d) Yes   11. (a) $g^{-1}(x) = \frac{1}{3}(x - 7)$

(b) $h^{-1}(x) = \sqrt[3]{\frac{1}{2}(x + 5)}$ (c) $f^{-1}(x) = \frac{3x}{1-x}, x \neq 1$

(d) $f^{-1}(x) = x^2 + 3, x \geq 0$

13. (a) Domain of f: all real numbers.   Domain of g: all real numbers except $2/3$.
(b) Range of f: all real numbers greater than or equal to -25.   (c) one-to-one

14. (a) $g^{-1}(x) = x^3 - 3$ (b) All real numbers

15. (a) $\frac{2}{x-1}$ (b) $\frac{x+5}{x-3}$ (c) $-2/3$ (d) -3 16. $5/4$

17. (a) $h^{-1}(x) = \frac{2}{x-1}, x \neq 1$ (b) $7/9$ 18. $p = 3, q = 2$

19. $p = 11/5, q = -2/5$

Answers to Exercises

Chapter 3

Exercise 3.1(a)

1. −5, −2   2. 3, 4   3. −5, 8   4. −5, 2   5. −5, −3
6. −3, 6   7. 4, 8   8. −7, 3   9. 2, 5   10. −4, 2
11. −6, 7   12. 3   13. −2, 6   14. −6, −5
15. −2, 10/3   16. 1, 3/2   17. −1/3, 1/2
18. −1/4, 3   19. 0, 1/4   20. −5, 5   21. −6, 2
22. −1, 7/3   23. −1/2, 3/2   24. −5, 8   25. −4, 7
26. −5, 6   27. 4, 6   28. −5/2, 3/2   29. −6, 2   30. 2/3, 2

Exercise 3.1(b)

1. (a) 36   (b) 49   (c) 16   (d) 64   (e) 9/4
   (f) 25/4   (g) 1/4   (h) 1/16   (i) 1/36
2. (a) −12.16, 0.16   (b) −0.65, 4.65   (c) −8.46, −1.54
   (d) −5.89, 2.89   (e) −6.91, 3.91   (f) 0.46, 6.54
   (g) −1.28, 0.78   (h) −1.26, 1.59   (i) −0.26, 1.26
   (j) −2.12, 0.12   (k) −1.69, 2.36   (l) − 0.53, 0.81

Exercise 3.1(c)

1. −13, 5   2. −2, 7   3. −1.45, 3.45   4. 0.46, 6.54
5. −10, 3   6. −1, 1/5   7. 3/4   8. 0.18, 2.82
9. −1, 1/3   10. − 0.28, 1.78   11. −1.46, 1.71
12. − 0.64, 0.52   13. − 0.93, 0.27   14. −3.7, 2.7
15. −1.9, 2.9   16. − 0.18, 1.85   17. − 0.15, 1.65
18. −1.45, 3.45   19. − 0.69, 2.19   20. −3.65, 1.65
21. −3.19, 0.52   22. − 0.61, 4.11

314   Answers to Exercises

Exercise 3.2

1. (a) Two distinct real roots (b) Imaginary roots (c) One real roots
   (d) Imaginary roots (e) Two distinct real roots (f) One real roots
   (g) Imaginary roots (h) Two distinct real roots

2. 1, 9   3. 4   4. $-1/3, 1$ 5. $k \leq 2$ 6. $k > -\dfrac{1}{4}$

7. 2   8. $k \leq 1, k \geq 5$ 9. 6   10. $k > 4$

Exercise 3.3

1. (a) $-17/4$ (b) -10 (c) 2
   (d) $-41/8$ (e) 1   (f) $1/2$

2. (a) $9/4$ (b) 6   (c) $49/4$
   (d) $49/8$ (e) $10/3$ (f) 4

3. (a) Minimum value = 1, $x = 3$
   (b) Maximum value = $25/4$, $x = -1/2$
   (c) Maximum value = 3, $x = -1$
   (d) Minimum value = 1, $x = -2$
   (e) Minimum value = 4, $x = -1$
   (f) Maximum value = 5, $x = 2$

Exercise 3.4

1. (a)   (b)   (c)

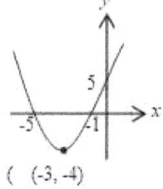
((-3, -4)

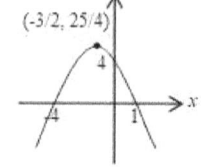
(-3/2, 25/4)

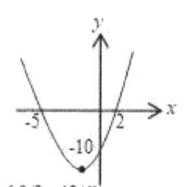
(-3/2, -48/4)

(d)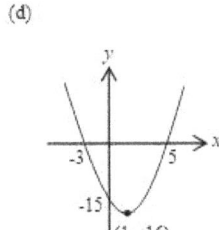

(e)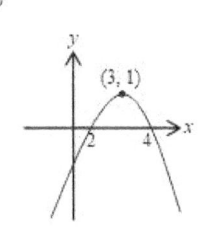

(f)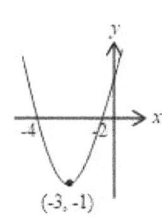

(g)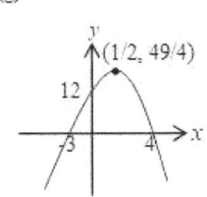

(h)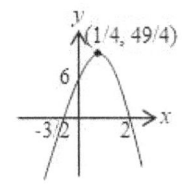

(i)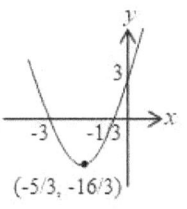

(j)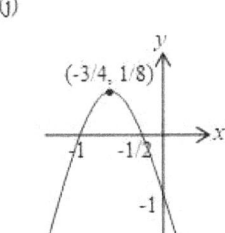

(k)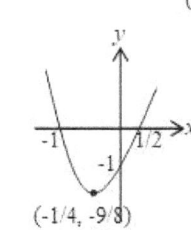

(l)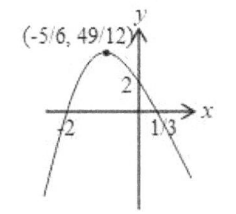

2. (a) All real numbers greater than or equal to -4

   (b) All real numbers less than or equal to 13

   (c) All real numbers greater than or equal to -2

   (d) All real numbers less than or equal to 21

   (e) All real numbers greater than or equal to $-25/4$

   (f) All real numbers greater than or equal to $-49/8$

   (g) All real numbers less than or equal to $25/4$

   (h) All real numbers greater than or equal to $-49/4$

Exercise 3.5(a)

1. $x < -1$ or $x > 4$
2. $-2 < x < 4$
3. $-3 \le x \le 4$
4. $x \le -4$ or $x \ge -3$
5. $-2 < x < 5$
6. $x < -9$ or $x > 3$
7. $-3 \le x \le 4$
8. $x < -3$ or $x > 6$
9. $-2 < x < -3/2$
10. $x \le -5/3$ or $x \ge -1$
11. $x \le -1$ or $x \ge 1/2$
12. $-3/2 \le x \le 3$

Exercise 3.5(b)

1. $-4 < x < 3$
2. $x < -5$ or $x > 2$
3. $x < -6$ or $x > 2$
4. $-5 < x < 3$
5. $x \le 2$ or $x \ge 3$
6. $-5 \le x \le -2$
7. $-2 \le x \le 3/2$
8. $-2/3 \le x \le 4$
9. $-1 < x < 3/4$
10. $x \le -8$ or $x \ge 3$

Review exercise 3

1. (a) $-5, 2$ (b) $-2, 10$ (c) $4, 5$ (d) $-2.32, 4.32$ (e) $-0.27, 1.47$ (f) $3, 7$
(g) $-1, 13$ (h) $-1.43, 2.10$ (i) $0.58, 2.42$ (j) $-3/2, 2$ (k) $-4, 3$ (l) $0.91, 1.66$

2. (a) One real roots (b) Two distinct real roots
(c) Imaginary roots (d) Two distinct real roots

3. $1, 5$ 4. $k < 3$ 5. 1 6. $k > 1/3$

7. (a) Minimum value $= -5/2$

(b) Maximum value $= -1$ (c) Minimum value $= 3$

(d) Minimum value $= -4$ (e) Maximum value $= 20$

(f) Minimum value $= -2$ (g) Maximum value $= 169/12$

(h) Minimum value $= -9/8$ (i) Maximum value $= 13/6$

8. (a) 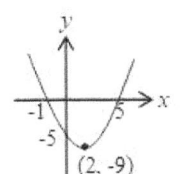   (b)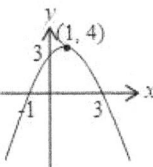

(c) 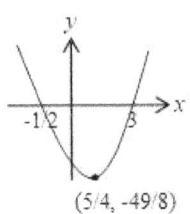   (d)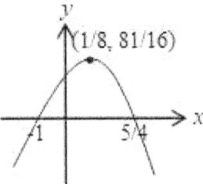

9. (a) All real numbers greater than or equal to -1

   (b) All real numbers less than or equal to $19/2$

   (c) All real numbers greater than or equal to 1

10. (a) $x \leq -5$ or $x \geq 1$ (b) $-1 < x < 7$ (c) $-5 \leq x \leq 2$

   (d) $2 < x < 6$ (e) $-2 \leq x \leq 4/3$ (f) $x \leq -5/2$ or $x \geq 4$

Chapter 4

Exercise 4.1(a)

1. (a) 4^5 (b) 3^{10} (c) 2^8 (d) 5^4 (e) 7^3

   (f) 4^2 (g) 5^6 (h) 3^{15} (i) 7^8

2. (a) 2^4 (b) 3^3 (c) 5^7

3. (a) $2a^4 b$ (b) $3x^3 y^2$ (c) $\frac{1}{3}x^2 y^4 z^2$

4. (a) $15a^5$ (b) $6a^5 b^3$ (c) $9x^4 y^6$

   (d) $72x^{12}$ (e) $72x^7 y^{11}$ (f) $4x^2 y^6$

Exercise 4.1(b)

1. (a) 6   (b) 5   (c) 3   (d) 8   (e) 16   (f) 16   (g) 1

   (h) $1/5$ (i) $1/5$ (j) $1/32$ (k) $1/8$ (l) 27

Answers to Exercises

2. (a) 8 (b) 9 (c) 8 (d) 9/16 (e) 27/8

  (f) 36/25 (g) 27/8 (h) 9/4 (i) 3/10

3. (a) 9/2 (b) 10 (c) 3 (d) 9

  (e) 15 (f) 1 (g) 16 (h) 2 (i) 5

4. (a) $\frac{x^2}{y^4}$ (b) 1 (c) $\frac{2yz^3}{x^2}$ (d) $\frac{x^3}{y^2 z}$ (e) $\frac{y}{x}$ (f) x^4

Exercise 4.1(c)

1. (a) 4/3 (b) 3/2 (c) 5/7 (d) 2 (e) –2 (f) 7/3

  (g) 5 (h) 81 (i) 1 (j) –3 (k) –1 (l) 0, 1/9

2. (a) 7 (b) 7/2 (c) 7/8 (d) 7/2

Exercise 4.2(a)

1. $2\sqrt{5}$ 2. $2\sqrt{7}$ 3. $3\sqrt{3}$ 4. $2\sqrt{10}$ 5. $3\sqrt{5}$ 6. $4\sqrt{3}$

7. $5\sqrt{2}$ 8. $3\sqrt{6}$ 9. $2\sqrt{15}$ 10. $6\sqrt{2}$ 11. $5\sqrt{3}$ 12. $4\sqrt{5}$

13. $4\sqrt{6}$ 14. $7\sqrt{2}$ 15. $7\sqrt{3}$ 16. $5\sqrt{6}$ 17. $9\sqrt{2}$ 18. $6\sqrt{5}$

19. $8\sqrt{3}$ 20. $7\sqrt{5}$

Exercise 4.2(b)

1. $7\sqrt{3}$ 2. $8\sqrt{2}$ 3. $7\sqrt{5}$ 4. $2\sqrt{7}$ 5. $-3\sqrt{3}$ 6. $-2\sqrt{5}$

7. $4\sqrt{3}$ 8. $8\sqrt{2}$ 9. $2\sqrt{3}$ 10. $2\sqrt{5}$ 11. $6\sqrt{2}$ 12. $6\sqrt{3}$

13. $4\sqrt{5}$ 14. $-3\sqrt{7}$ 15. $7\sqrt{2}$ 16. $3\sqrt{2}$

Exercise 4.2(c)

1. $2\sqrt{3}$ 2. $5\sqrt{3}$ 3. $10\sqrt{15}$ 4. $6\sqrt{10}$

5. $4\sqrt{5}$ 6. $9\sqrt{5}$ 7. $10\sqrt{15}$ 8. $35\sqrt{6}$

9. $2\sqrt{15}$ 10. $12\sqrt{5}$ 11. $18\sqrt{5}$ 12. $180\sqrt{3}$

Exercise 4.2(d)

1. $\frac{2\sqrt{3}}{3}$ 2. $2\sqrt{5}$ 3. $3\sqrt{3}$ 4. $\frac{3\sqrt{10}}{2}$ 5. $3\sqrt{3}$ 6. $2\sqrt{2}$

7. $\frac{\sqrt{30}}{2}$ 8. $4\sqrt{3}$ 9. $\sqrt{2}$ 10. $\frac{3\sqrt{2}}{8}$ 11. $\frac{\sqrt{10}}{6}$ 12. $\frac{2\sqrt{7}}{5}$

Exercise 4.2(e)

1. $\sqrt{15} - 3$ 2. $3\sqrt{30} - 4\sqrt{6}$ 3. $6 + \sqrt{30}$ 4. $2\sqrt{21} + 21$

5. $13 + 7\sqrt{3}$ 6. $7 - 3\sqrt{5}$ 7. $5\sqrt{6}$ 8. $2\sqrt{6} - 2$ 9. 4

10. -20 11. $22 + 12\sqrt{2}$ 12. $30 - 12\sqrt{6}$

Exercise 4.2(f)

1. $\sqrt{2} - 1$ 2. $2 + \sqrt{3}$ 3. $\frac{4-\sqrt{6}}{2}$ 4. $\frac{5\sqrt{7}+7}{6}$ 5. $\frac{12\sqrt{2}-2\sqrt{6}}{11}$

6. $\frac{12-4\sqrt{15}}{7}$ 7. $17 - 12\sqrt{2}$ 8. $-4 - \sqrt{15}$ 9. $\frac{1-2\sqrt{14}}{5}$

10. $-7 - 4\sqrt{3}$ 11. $\frac{5\sqrt{5}+11}{-2}$ 12. $\frac{2\sqrt{6}+3\sqrt{2}}{3}$

Review exercise 4

1. (a) 5^6 (b) 3^3 (c) 2^8 (d) 7^3 (e) 5^5
   (f) 3^2 (g) 2^8 (h) 3^6 (i) 5^6

2. (a) 3^3 (b) 7^3 (c) 5

3. (a) $\frac{3a^2b^2}{2}$ (b) $5x^4y^2$ (c) $\frac{3z^8y^2}{2}$

4. (a) $16x^8$ (b) $6x^3y^5$ (c) $15xy^2$

5. (a) $\frac{y^4z^8}{x^2}$ (b) $\frac{1}{xy^5}$ (c) $\frac{x}{y^3z^4}$

6. (a) $4/3$ (b) 125

7. (a) 10 (b) $2/3$ (c) $4/3$ (d) 0 (e) 2 (f) -3

8. (a) $-5/2$ (b) $-5/29$

Answers to Exercises

9. (a) $-5\sqrt{6}$ (b) $4\sqrt{2}$ (c) $-6\sqrt{5}$ (d) $23 + 7\sqrt{6}$

(e) $79 - 20\sqrt{3}$ (f) 10

10. (a) $\frac{\sqrt{2}}{10}$ (b) $\frac{4\sqrt{2}}{3}$ (c) $\frac{4\sqrt{2}}{9}$ (d) $\frac{3\sqrt{6}}{8}$ (e) $-12 - 9\sqrt{2}$

(f) $\frac{8+7\sqrt{3}}{11}$ (g) $5 + 2\sqrt{6}$ (h) $\frac{\sqrt{6}-5\sqrt{2}}{4}$

Chapter 5

Exercise 5.1

1. (a) -13 (b) 2 (c) 0 (d) -50 (e) 1

2. (a) $x^2 + 3x + 6, 7$ (b) $x^2 - x + 2, 2$ (c) $x^2 + 2x - 3, 10$

(d) $3x^2 + x + 6, 4$ (e) $x^2 - 3x + 4, 0$ (f) $x^2 + 2x + 2, -10$

Exercise 5.2

1. (a) 2 (b) 0 (c) 31 (d) -16 (e) 4 (f) $5/2$

2. (a) $(x - 3)(x - 1)(x + 2)$ (b) $(x + 1)(x + 2)(x + 3)$

(c) $(x - 2)(x + 3)(x + 4)$ (d) $(x - 2)(x + 3)(2x + 1)$

3. (a) $(x - 3)(x - 1)(x + 1)$ (b) $(x - 2)(x + 1)(x + 3)$

(c) $(2x - 1)(x - 2)(x + 1)$ (d) $(x - 3)(x + 3)(2x + 1)$

4. 4 5. -4 6. $-1/2, 1$ 7. $p = 1, q = -3$ 8. $a = 2, b = -10, 2x + 5$

9. $a = -5, b = -3, (x - 3)(x + 1)^2$ 10. $a = 2, b = -5$ 11. $p = 1, q = 2$

12. $a = 1, b = -3$

Exercise 5.3

1. (a) $-2, 1, 3$ (b) $-2, 3, 4$ (c) $-3, -2, 2$

(d) $-2, 1, 5$ (e) $-4, -1, 3$ (f) $-4, -1/2, 3$

2. (a) $-3, -1, 2$ (b) $-3, -2, -1$ (c) $-2, -1/2, 1$

(d) $-2, -1/3, 1$

Review exercise 5

1. (a) $x^2 + x + 2$ (b) $x^2 - 1$ (c) $2x^2 - x + 3$ (d) $x^2 - 1$

2. (a) 18 (b) 0 (c) 0 (d) 2

3. (a) $(x - 4)(x - 2)(x + 1)$ (b) $(3x - 5)(x - 2)(x + 1)$
   (c) $(2x - 1)(x - 2)(x + 2)$ (d) $(x - 2)(x - 1)(x + 3)$

4. $(x - 1), (x - 2)$

5. (a) $(x - 3)(x - 1)(x + 2)$; $-2, 1, 3$ (b) $(x - 3)(x - 2)(x + 1)$; $-1, 2, 3$
   (c) $(x - 3)(x - 2)(x - 1)$; $1, 2, 3$ (d) $(2x - 1)(x - 2)(x + 2)$; $-2, 1/2, 2$
   (e) $(2x - 3)(x - 2)^2$; $3/2, 2$ (f) $(x - 5)(x - 3)(x - 1)(x + 1)$; $-1, 1, 3, 5$

6. $a = 3, b = 7, c = -6, p = -20$ $(3x - 2)(x - 2)(x + 3)$

7. $a = 2, b = -5, c = 3, p = 3$ $(2x - 3)(x - 1)(x + 1)$

8. $p = -2$; $-1, 1, 1/2$ 9. $a = -7, b = -27$ 10. 30 11. 2/3, 2

12. (a) $(3x - 4)(x - 2)(x + 2)$; $-2, 4/3, 2$
    (b) $(2x - 3)(2x + 1)(x + 1)$; $-1, -1/2, 3/2$
    (c) $(x - 1)(4x + 3)(x + 2)$; $-2, -3/4, 1$
    (d) $(x - 2)(3x + 2)(2x + 1)$; $-1/2, -2/3, 2$

Chapter 6

Exercise 6.1(a)

1. $x = 2, y = 1$ 2. $x = 3, y = 2$ 3. $x = 4, y = 3/2$
4. $x = -8, y = 16$ 5. $x = 2, y = 2$ 6. $x = 5, y = -11/2$
7. $x = -1, y = 0$ 8. $x = 3, y = 4$ 9. $x = 5/2, y = 1$
10. $x = 3, y = 4/3$ 11. $x = 2, y = -4$ 12. $x = -2, y = 3$

Exercise 6.1(b)

1. $x = 1, y = 2$ 2. $x = 3, y = 2$ 3. $x = 3, y = 1$

4. $x = -1, y = -1$ 5. $x = 2, y = 2$ 6. $x = 0, y = 3$

7. $x = 0, y = -2$ 8. $x = -1, y = 6$ 9. $x = 1, y = -3$

10. $x = -2, y = 3$ 11. $x = -2, y = -3$ 12. $x = 2, y = 2$

Exercise 6.2

1. $x = 2, y = 1; x = 8, y = -11$ 2. $x = 3, y = 1; x = 27/11, y = 29/11$

3. $x = -1, y = -4; x = 5, y = 8$ 4. $x = -8, y = -3; x = 4, y = 1$

5. $x = 2, y = 3; x = -7/5, y = -19/5$ 6. $x = 2, y = -1; x = 6, y = 5$

7. $x = 3/4, y = 3/2; x = 3, y = -3$ 8. $x = 3, y = 4$ 9. $x = 4, y = -1$

Review exercise 6

1. $x = 6, y = -9$ 2. $x = 2, y = -1$ 3. $a = 4, b = 4$

4. $x = 2, y = -1$ 5. $a = 5, b = 5$ 6. $x = -1, y = 2$

7. $x = 2, y = 3$ 8. $x = 8, y = 7$ 9. $x = 2, y = 7$

10. $x = 4, y = 3$ 11. $x = 3, y = 4$ 12. $x = -2, y = -3$

13. $x = 12, y = 18$ 14. $a = 4, b = 3$ 15. $x = -1, y = 3; x = -3, y = 1$

16. $x = 0, y = 2; x = 2, y = 0$ 17. $x = -1, y = -4; x = 3, y = 0$

18. $x = -1/3, y = 3/2; x = 1, y = 0$ 19. $x = -2, y = 4; x = 4 y = -2$

20. $x = -3, y = 0; x = 1, y = 2$

Chapter 7

Exercise 7(a)

1. 5   2. 3   3. –7   4. 3   5. –2   6. 1/2   7. 6   8. –2

Exercise 7(b)

1. $3 \log_a x + 2 \log_a y$ 2. $2 \log_a x - \log_a y$ 3. $2 \log x + \frac{1}{3} \log y$

4. $-3 \log_2 x$ 5. $3 \ln x + \frac{1}{2} \ln y$ 6. $2 \log_3 x + 3 \log_3 y - 2 \log_3 z$

7. $2 \log_5 x + \log_5 y - \frac{1}{3} \log_5 z$ 8. $4 \ln x + \frac{1}{3} \ln y - 3 \ln z$

Answers to Exercises 323

9. $\frac{1}{4}\log x + \frac{1}{2}\log y - \frac{1}{2}\log z$ 10. $\ln x + 2\ln y - \frac{4}{3}\ln z$

11. $2\log_2 x + 3\log_2 y - \frac{5}{2}\log_2 z$ 12. $1 + \frac{2}{3}\log x - \log y - \frac{4}{3}\log z$

Exercise 7(c)

1. $\log_2 x^2 y^3$ 2. $\log_3 \left(\frac{x^2}{y^3 z}\right)$ 3. $\log_3 xy^2$ 4. $\log_5 \left(\frac{\sqrt{x}}{\sqrt[3]{y^2 z^3}}\right)$

5. $\log \left(\frac{x^3 z^6}{y^9}\right)$ 6. $\log x^3 y$ 7. $\log \sqrt[3]{x^2} y^2$ 8. $\ln \left(\frac{\sqrt{x}\sqrt[4]{y^3}}{z}\right)$

9. $\log x \sqrt[3]{y^2}$ 10. $\ln \left(\frac{x^3 y^2}{z}\right)$ 11. $\log \left(\frac{1000 x^2}{\sqrt[4]{y}}\right)$ 12. $\log \left(\frac{10\sqrt[3]{x^2}}{y^3}\right)$

Exercise 7(d)

1. 5   2. 7   3. 3   4. 3   5. −2   6. 3   7. 4   8. −3

9. −1/3   10. 1/3   11. 4/3   12. 2

Exercise 7(e)

1. 4.7   2. 1.585   3. 1.431   4. 1.893   5. 1.277   6. 1.511

7. 1.667   8. 0.946   9. 1.285   10. 1.723   11. 1.850   12. 1.674

Exercise 7(f)

1. 1.292   2. 2.161   3. 0.631   4. 0.670   5. 1.453   6. 3.819   7. 3.226

8. −2.369   9. −0.639   10. 0.118   11. −4   12. 1.708

Exercise 7(g)

1. 7   2. 1   3. 8   4. 3/2   5. 13   6. 6   7. 20

8. 19/8   9. 12/19   10. 2   11. 19/4   12. 1/3, 9

Review exercise 7

1. (a) $2\log x + \log y - 3\log z$ (b) $x \ln x + 3\ln y - 2\ln z$

(c) $4 + 2\log_2 x + 3\log_2 y - 4\log_2 z$ (d) $2\ln x + 3\ln y - 4\ln z$

(e) $\frac{5}{2}\log_5 x - \frac{9}{2}\log_5 y - \frac{7}{2}\log_5 z$ (f) $2\log x + \frac{2}{3}\log y - \frac{4}{3}\log z$

2. (a) $\log x^3 y^5$ (b) $\log\left(\frac{x^2}{y^3}\right)$ (c) $\ln\left(\frac{x^5 y^3}{z^2}\right)$ (d) $\log_2\left(\frac{\sqrt[5]{x^3}}{\sqrt[3]{y^2}}\right)$

(e) $\log\left(\frac{\sqrt[4]{x}\sqrt{y^3}}{z^3}\right)$ (f) $\ln x^2 \sqrt[4]{y}$ (g) $\ln\left(\frac{\sqrt[3]{xz^2}}{y^3}\right)$ (h) $\log_2\left(\frac{x^3}{256 y^2}\right)$

3. (a) 4.106 (b) 1.292 (c) 1.608 (d) 0.734

4. (a) 9 (b) 3 (c) 5 (d) 36 (e) 2 (f) 2

Chapter 8

Exercise 8.1

1. (a) 13 (b) 5 (c) 8 (d) $\sqrt{13}$ (e) $\sqrt{13}$ (f) $2\sqrt{2}$

2. $AB = 3\sqrt{2}$, $AC = 2\sqrt{17}$, $BC = 5\sqrt{2}$ ABC is a right angled triangle

3. The triangle is isosceles 4. 3 5. $-4, 2$ 6. 4

Exercise 8.2

1. (a) (3, 3) (b) (5, 6) (c) $(3, -1/2)$ (d) $(4, -1)$

(e) $(-5/2, -3/2)$ (f) (2, 2)

2. (5, 3), $(-3/2, -3/2)$, $(1/2, -1/2)$ 3. (4, 8) 4. $x = 9, y = 5$

Exercise 8.3

1. (a) $7/4$ (b) -2 (c) $1/11$ (d) $3/2$ (e) $-5/6$

3. (a) collinear (b) not collinear (c) not collinear (d) collinear

4. 14 5. 5 6. 2 7. 1 8. $a = 2, b = 1$

Exercise 8.4

1. $y = 2x - 2$ 2. $y = 3x - 1$ 3. $y = 4x - 11$ 4. $y = -3x + 13$

5. $y = 3x + 14$ 6. $3x + 2y + 2 = 0$ 7. $4x + 3y - 6 = 0$

8. $3x + 4y + 22 = 0$ 9. $5x + 4y - 32 = 0$ 10. $3x - 4y = 0$

11. $2x - 3y - 16 = 0$ 12. $3x - 2y + 5 = 0$

Answers to Exercises

Exercise 8.5

1. (a) $4x - 3y - 7 = 0$ (b) $5x - 2y + 13 = 0$
   (c) $2x + 3y = 0$ (d) $3x - 2y + 4 = 0$

2. (a) $3x - 5y + 3 = 0$ (b) $y = 3x + 11$
   (c) $2x - 3y + 12 = 0$ (d) $3x + 2y - 1 = 0$
   (e) $2x - 3y + 4 = 0$ (f) $3x + 2y - 12 = 0$

3. 11   4. 3   5. $6x + 4y - 29 = 0$ 6. $y = -x + 7$

7. -1 8. $9/4$ 9. $3x + 2y - 12 = 0$

10. ABC is a right angled triangle   6.5

Exercise 8.6

1. (a) $(3, 2)$ (b) $(3, 3)$ (c) $(5, -11/2)$ (d) $(2, -4)$
   (e) $(2, 2)$ 2. $(3, 9)$ 3. $(-4, 5)$ 4. $4/3$

5. $(1, 3)$ 6. $(1, 10)$ 7. $P(-1, 0)$, $Q(0, 4/3)$ $2/3$

Exercise 8.7

1. $a = 5.6, n = 2$ 2. $a = 3, b = 2$ 3. $a = 15, n = -2, 27$
4. $y = 4.2x^{1.6}$ 5. $a = 1.5, b = 8, x = 18$ 6. $a = -1.2, b = 7$
7. $a = 3, b = 2.4$ 8. $a = 0.45, b = 1.2$ 9. $a = 4.8, b = -0.5$

Review exercise 8

1. (a) $\sqrt{5}$ (b) $\sqrt{10}$ (c) 5 (d) $\sqrt{13}$ (e) $\sqrt{29}$ (f) 13

2. (a) $(2, 5)$ (b) $(3, 1)$ (c) $(9/2, 11/2)$ (d) $(13/2, 4)$
   (e) $(13/2, 5/2)$ (f) $(-2, 7/2)$

3. (a) $(9/2, 7/2)$ (b) $(-5/2, -4)$ (c) $(11/2, -2)$
   (d) $(9/2, -4)$ (e) $(1/2, 3)$ (f) $(1/2, 5/2)$

4. (a) $x - 2y + 7 = 0$ (b) $5x + 2y - 22 = 0$ (c) $4x - 3y - 2 = 0$

(d) $4x - 3y + 20 = 0$ (e) $3x + 4y - 1 = 0$ (f) $y = 3x - 13$

5. (a) $x - 4y + 5 = 0$ (b) $2x + 5y - 1 = 0$

(c) $3x - 2y - 8 = 0$ (d) $3x + 2y - 17 = 0$

6. (a) $2x - 3y + 12 = 0$ (b) $y = -4x + 11$

(c) $y = -3x - 9$ (d) $3x - 4y - 11 = 0$

7. (a) (3, −4)   (b) (7, −2)   (c) (−1, −1)   (d) (2, −5)

11. (a) $(-7/2, 1/2)$, $(3/2, -3/2)$, $(-3/2, 7/2)$, $(-1/2, -9/2)$

12. $y = 3x - 6$ 13. $x - 3y - 6 = 0$

14. $7x + 8y - 17 = 0$ 15. (a) $a = 3, b = 1.5$ (b) 18.6

16. $a = 1.6, n = 2$ 17. (a) $a = 0.5, b = 4.2$ (b) 17.0

Chapter 9

Exercise 9.1

1. (a) $\pi/6$ (b) $\pi/3$ (c) $\pi/2$ (d) $2\pi/3$ (e) $5\pi/6$ (f) $-7\pi/6$

(g) $5\pi/4$ (h) $3\pi/2$ (i) $9\pi/4$ (j) $-4\pi/3$ (k) $-11\pi/6$ (l) $5\pi/3$

2. (a) 36°   (b) 135°   (c) 420°   (d) 540°   (e) −225°   (f) 144°

(g) −252°   (h) −480°   (i) 80°   (j) 18°   (k) −720°   (l) 450°

Exercise 9.2

1. (a) 10.8 cm, 81 cm^2 (b) 20 m, 80 m^2 (c) 7.2 m, 21.6 m^2

(d) 25.13 cm, 150.8 cm^2 (e) 94.25 cm, 942.48 cm^2 (f) 39.27 m, 294.52 m^2

2. (a) 1.2 rad.   (b) 2.5 rad.   (c) 2.1 rad.   (d) 0.8 rad.

3. 4 rad.   4. 6 cm   5. 4.8 cm   6. 7 cm   7. 6.75 cm^2

8. 19.4 cm   9. 45.4 cm   10. 57.4 cm   11. 36.18 cm^2, 416.2 cm^2

12. 37.07 cm^2 13. 7.8 cm   14. 30.15 cm^2, 34.9 cm

15. 2 : 25   16. 20.94 cm, 61.42 cm^2

Review exercise 9

1. (a) $11\pi/4$ (b) $-\pi/4$ (c) $-7\pi/3$ (d) $10\pi/3$
   (e) $-5\pi/3$ (f) $9\pi/5$

2. (a) $60°$ (b) $270°$ (c) $-45°$ (d) $900°$
   (e) $288°$ (f) $156°$

3. (a) 27 cm, 121.5 cm^2 (b) 37.5 m, 281.25 m^2
   (c) 20.03 cm, 85.12 cm^2 (d) 30.16 m, 180.96 m^2

4. (a) 0.6 rad.   (b) 2.5 rad.   (c) 2.4 rad.   (d) 0.2 rad.

5. 3 cm   6. 24π 7. 1/22 s   8. 937.5 m^2 9. (a) 3.96 cm^2, 42.2%

10. (a) 1.8 rad.   (b) 22 cm   (c) 18 cm^2 11. (a) 80.9 cm^2 (b) 43.67 cm

12. (a) 31.5 cm   (b) 97.11 cm^2 (c) 567.72 cm^2

13. (a) OD = 2 m   OC = 1.5 cm   (b) 14.8 cm   (c) 13.3 cm^2

14. 3.26 cm^2 15. 34.36 cm

Chapter 10

Exercise 10.1

1. (a) $-\sqrt{3}/2$ (b) $-\sqrt{3}$ (c) $-1/2$ (d) 1   (e) $\sqrt{2}/2$
   (f) $-\sqrt{2}$ (g) $2\sqrt{3}/3$ (h) $-2\sqrt{3}/3$ (i) 1   (j) $-\sqrt{3}$
   (k) $\sqrt{2}/2$ (l) $1/2$ (m) -2 (n) 1   (o) 0

2. (a) 0   (b) $\sqrt{2}/2$ (c) $\sqrt{3}/3$ (d) $2\sqrt{3}/3$ (e) $-2\sqrt{3}/3$
   (f) -1 (g) $-\sqrt{3}$ (h) -1 (i) $-1/2$ (j) 2   (k) $-\sqrt{3}/3$
   (l) $\sqrt{2}/2$ (m) $-1/2$ (n) $-\sqrt{2}/2$ (o) $\sqrt{3}$

Exercise 10.2(a)

1. $\cos\theta$ 2. $\csc\theta$ 3. $\tan^2\theta$ 4. $\sec^2\theta$

5. $1 + \cos\theta$ 6. $\csc\theta + 1$ 7. $\tan^2\theta$ 8. $\tan\theta$

328 Answers to Exercises

9. 1   10. 2   11. $\operatorname{cosec} \theta$ 12. $2 \sin \theta \cos \theta$

Exercise 10.3

1. (a) 45°, 135°, 225°, 315°   (b) 30°, 150°, 210°, 330°   (c) 30°, 210°

(d) 60°, 120°, 240°, 300°   (e) 60°, 120°, 240°, 300°   (f) 45°, 225°

(g) 60°, 120°, 240°, 300°   (h) 60°, 120°, 240°, 300°

(i) 60°, 300°   (j) 120°, 240°

2. (a) $\pi/3, \pi, 5\pi/3$ (b) $\pi/3, 2\pi/3, 4\pi/3, 5\pi/3$ (c) $\pi/4, \pi, 5\pi/4$

(d) $3\pi/2$ (e) $2\pi/3, 4\pi/3$ (f) $7\pi/6, 3\pi/2, 11\pi/6$

Exercise 10.4

1. (a) Amplitude = 3, Period = $\pi/2$ (b) Amplitude = 2, Period = $2\pi/3$

(c) Amplitude = 2, Period = $2\pi/5$ (d) Amplitude = 5, Period = 4π

(e) Amplitude = 1, Period = $\pi/3$ (f) Amplitude = 2, Period = $8\pi/3$

Review exercise 10

1. (a) $-\pi/4$ (b) $-\pi/3$ (c) $-\pi/6$ (d) $\sqrt{3}/3$

(e) $1/2$ (f) $-\sqrt{2}/2$

2. (a) $\cos^2 \theta$ (b) $\cot^2 \theta$ (c) 2   (d) 1   (e) $\operatorname{cosec} \theta$ (f) $2 \sec \theta$

4. (a) $\pi/4, 3\pi/4, 5\pi/4, 7\pi/4$ (b) $\pi/3, 2\pi/3, 4\pi/3, 5\pi/3$

(c) $\pi/3, 2\pi/3, 4\pi/3, 5\pi/3$ (d) $\pi/6, 5\pi/6, 7\pi/6, 11\pi/6$

5. (a) Amplitude = 5, Period = π (b) Amplitude = 3, Period = $\pi/3$

(c) Amplitude = 2, Period = $2\pi/3$ (d) Amplitude = 2, Period = $\pi/2$

(e) Amplitude = 4, Period = 4π (f) Amplitude = 6, Period = 3π

Chapter 11

Exercise 11.1

1. 20   2. 18   3. 864   4. 1320   5. 360   6. 5040   7. 36

8. 544320   9. 20000   10. 256   11. 60   12. 50

Exercise 11.2

1. 720   2. 720   3. 40320   4. 6840   5. 90   6. 5040

7. 524160   8. 5040   9. 120   10. 151200   11. 720

12. 13800   13. 4536   14. 24360   15. 20160   16. 6720

17. (a) 60   (b) 125   18. 729   19. 250   20. 45

21. 72   22. 18   23. 120   24. 240   25. 36

Exercise 11.3

1. 56   2. 126   3. 210   4. 2300   5. 495   6. 18

7. 60   8. 6435   9. 4845   10. 36   11. 220   12. 35

13. 70   14. 120   15. 364   16. 20   17. 1260   18. 1800

19. 150   20. 110   21. 140   22. 40   23. 165

24. (a) 1716   (b) 140   (c) 1008

25. (a) 6435   (b) 1890   (c) 540

Review exercise 11

1. 120   2. 720   3. 110   4. 210   5. 6720   6. 24

7. 182   8. 288   9. 210   10. 336, 56   11. 10

12. (a) 15,120   (b) 59,049   13. (a) 300   (b) 144   14. 240

15. (a) 1296   (b) 96   16. (a) 3003   (b) 1260   (c) 861

17. (a) 924   (b) 28   (c) 672

Chapter 12

Exercise 12.1

1. (a) 84   (b) 1   (c) 10   (d) 15   (e) 35   (f) 252   (g) 220   (h) 6435

2. (a) $240x^2$ (b) $-22680x^4$ (c) $-63x^5/8$ (d) $2160x^4$

3. (a) 7920   (b) 5376   (c) 1180980   (d) $945/16$

4. (a) $x^4 + 4x^3y + 6x^2y^2 + 4xy^3 + y^4$

   (b) $32x^5 - 80y^4y + 80x^3y^2 - 40x^2y^3 + 10xy^4 - y^5$

   (c) $81a^4 + 216a^3b + 216a^2b^2 + 96ab^3 + 16b^4$

   (d) $64x^3 - 144x^2y + 108xy^2 - 27y^3$

5. (a) $1 - 10x + 45x^2 - 120x^3$ (b) $1 + 18x + 144x^2 + 672x^3$

   (c) $1024 - 5120x + 11520x^2 - 15360x^3$ (d) $1 + 3x + \frac{15}{4}x^2 + \frac{5}{2}x^3$

   (e) $2187 - 3402x + 2268x^2 - 840x^3$ (f) $256 + 512x + 448x^2 + 224x^3$

6. 240   7. $40x$ 8. $70x^4/y^4$

9. (a) $1 + 7x + 14x^2$ (b) $1 - 12x + 78x^2$

   (c) $1 - 6x + 14x^2 - 14x^3$ (d) $1 + 9x + 30x^2 + 40x^3$

10. $1 - 20x + 180x^2 - 960x^3$, 08170   11. $1 + \frac{6}{5}x + \frac{3}{5}x^2 + \frac{4}{25}x^3$, 1.062

12. $n = 5, a = 2$ 13. 3

Exercise 12.2

1. (a) $1 + 12x + 60x^2 + 160x^3 + 240x^4 + 192x^5 + 64x^6$

   (b) $x^4 - 8x^3y + 24x^2y^2 - 32xy^3 + 16y^4$

   (c) $1 - 5x + 10x^2 - 10x^3 + 5x^4 - x^5$

   (d) $16x^4 + 96x^3y + 216x^2y^2 + 216xy^3 + 81y^4$

   (e) $64x^3 - 48x^2 + 12x - 1$

   (f) $81x^4 + 108x^3y + 54x^2y^2 + 12xy^3 + y^4$

   (g) $1 - 12x + 54x^2 - 108x^3 + 81x^4$

   (h) $64a^6 + 192a^5x + 240a^4x^2 + 160a^3x^3 + 60a^2x^4 + 12ax^5 + x^6$

   (i) $16a^4 - 96a^3x + 216a^2x^2 - 216ax^3 + 81x^4$

2. $1 - 5x + 10x^2 - 10x^3 + 5x^4 - x^5$, 0.904

3. $1 + 4x + 6x^2 + 4x^3 + x^4$, 1.0406

4. $1 - 10x + 40x^2 - 80x^3 + 80x^4 - 32x^5$, 0.904

Review exercise 12

1. (a) 792 (b) 210 (c) 5005 (d) 12870 (e) 1287 (f) 12650

2. (a) $x^3 + 6x^2 + 12x + 8$

   (b) $x^5 + 15x^4 + 90x^3 + 270x^2 + 405x + 243$

   (c) $x^4 - 4x^3y + 6x^2y^2 - 4xy^3 + y^4$

   (d) $32x^5 - 80x^4 + 80x^3 - 40x^2 + 10x - 1$

   (e) $64x^6 + 192x^5y + 240x^4y^2 + 160x^3y^3 + 60x^2y^4 + 12xy^5 + y^6$

   (f) $1 - 8x + 24x^2 - 32x^3 + 16x^4$

3. (a) $x^4 - 12x^3 + 54x^2 - 108x + 81$

   (b) $x^3 + 12x^2 + 48x + 64$

   (c) $8x^3 - 12x^2y + 6xy^2 - y^3$

   (d) $32x^5 + 80x^4y + 80x^3y^2 + 40x^2y^3 + 10xy^4 + y^5$

   (e) $81x^4 + 216x^3y + 216x^2y^2 + 96xy^3 + 16y^4$

   (f) $64x^3 - 144x^2y + 108xy^2 - 27y^3$

4. (a) $512 + 2304x + 4608x^2 + 5376x^3$

   (b) $2187 - 5103x + 5103x^2 - 2835x^3$

   (c) $1 + 24x + 264x^2 + 1760x^3$

   (d) $1 - 24x + 252x^2 - 1512x^3$

   (e) $1 + \frac{7}{2}x + \frac{21}{4}x^2 + \frac{35}{8}x^3$

   (f) $1 - 9x + \frac{135}{4}x^2 - \frac{135}{2}x^3$

5. (a) $-120x^7y^3$ (b) $5940x^3y^9$ (c) $5103x^5y^2$ (d) $-129024a^4b^5$

6. (a) -120 (b) 1365 (c) -1760 (d) 120 (e) 54 (f) -5940

7. (a) 240 (b) 7/16 (c) 11520

8. $x^4 + 4x^3y + 6x^2y^2 + 4xy^3 + y^4$, 15.682

9. (a) $1 + 7x + 21x^2 + 35x^3 + 35x^4 + 21x^5 + 7x^6 + x^7$ (b) 0.97919

10. (a) $1 - \frac{8}{3}x + \frac{28}{9}x^2 - \frac{56}{27}x^3 + \frac{70}{81}x^4$ (b) 0.97625

11. $1 - 12x + 60x^2 - 160x^3 + 240x^4 - 192x^5 + 64x^6$, 0.690

12. ±4   13. ±3

14. (a) $192 + 80x - 3x^2 - \frac{15}{2}x^3$ (b) $1 - \frac{23}{2}x + 54x^2 - 130x^3$

(c) $256 - 512x - 256x^2 + 1792x^3$ (d) $64 + 64x - 80x^2 - 128x^3$

Chapter 13

Exercise 13.1(a)

1. (a) $\begin{pmatrix} 3 \\ -1 \end{pmatrix}$ (b) $\begin{pmatrix} -1 \\ 2 \end{pmatrix}$ (c) $\begin{pmatrix} 0 \\ 1 \end{pmatrix}$

(d) $\begin{pmatrix} -1 \\ -1 \end{pmatrix}$ (e) $\begin{pmatrix} -6 \\ 7 \end{pmatrix}$ (f) $\begin{pmatrix} 3 \\ -2 \end{pmatrix}$

2. (a) $x = 5, y = 5$ (b) $x = -3, y = 2$

(c) $x = 1, y = 2$ (d) $x = 2, y = -1$

3. $x = -3, y = 4$

Exercise 13.1(b)

1. (a) $\begin{pmatrix} -1 \\ -3 \end{pmatrix}$ (b) $\begin{pmatrix} -3 \\ 4 \end{pmatrix}$ (c) $\begin{pmatrix} 2 \\ -5 \end{pmatrix}$ (d) $\begin{pmatrix} 2 \\ 0 \end{pmatrix}$

2. (a) $\begin{pmatrix} -2 \\ -3 \end{pmatrix}$ (b) $\begin{pmatrix} 2 \\ -4 \end{pmatrix}$ (c) $\begin{pmatrix} -6 \\ 3 \end{pmatrix}$ (d) $\begin{pmatrix} 4 \\ -1 \end{pmatrix}$

Exercise 13.1(c)

1. (a) $\begin{pmatrix} 2 \\ 3 \end{pmatrix}$ (b) $\begin{pmatrix} 7 \\ -1 \end{pmatrix}$ (c) $\begin{pmatrix} 0 \\ 4 \end{pmatrix}$

(d) $\begin{pmatrix} -2 \\ -9 \end{pmatrix}$ (e) $\begin{pmatrix} -3 \\ -1 \end{pmatrix}$ (f) $\begin{pmatrix} 6 \\ 5 \end{pmatrix}$

2. (a) $\begin{pmatrix} 1 \\ -1 \end{pmatrix}$ (b) $\begin{pmatrix} 2 \\ 4 \end{pmatrix}$ (c) $\begin{pmatrix} -1 \\ -5 \end{pmatrix}$

Exercise 13.1(d)

1. (a) $\begin{pmatrix} 6 \\ -2 \end{pmatrix}$ (b) $\begin{pmatrix} 3 \\ -6 \end{pmatrix}$ (c) $\begin{pmatrix} -1 \\ 3 \end{pmatrix}$

 (d) $\begin{pmatrix} -2 \\ 4 \end{pmatrix}$ (e) $\begin{pmatrix} 6 \\ 0 \end{pmatrix}$ (f) $\begin{pmatrix} -4 \\ -8 \end{pmatrix}$

2. (a) $\begin{pmatrix} -2 \\ -4 \end{pmatrix}$ (b) $\begin{pmatrix} 6 \\ -9 \end{pmatrix}$ (c) $\begin{pmatrix} 7 \\ 0 \end{pmatrix}$ (d) $\begin{pmatrix} 1 \\ -10 \end{pmatrix}$

3. $x = 3, y = 2$ 4. $x = 1, y = 1$ 5. $k = 2, m = 3$

Exercise 13.1(e)

1. (a) 5   (b) $2\sqrt{5}$ (c) $\sqrt{34}$ (d) 13

2. (a) $\sqrt{85}$ (b) $\sqrt{73}$ (c) $5\sqrt{10}$ (d) $\sqrt{145}$

Exercise 13.1(f)

1. (a) $4i + 3j$ (b) $3i - 2j$ (c) $-5i - 2j$

 d) $5j$ (e) $-7i$

2. (a) $-i - j$ (b) $-3i - 7j$ (c) $12i + 5j$ (d) $-13i + 12j$

3. (a) $\sqrt{10}$ (b) 5   (c) 12   (d) 17

4. (a) $\frac{\sqrt{2}}{2}i + \frac{\sqrt{2}}{2}j$ (b) $\frac{-3\sqrt{13}}{13}i + \frac{2\sqrt{13}}{13}j$

 (c) $\frac{2\sqrt{5}}{5}i - \frac{\sqrt{5}}{5}j$ (d) $\frac{-\sqrt{2}}{10}i - \frac{7\sqrt{2}}{10}j$

Exercise 13.1(g)

1. (a) $\begin{pmatrix} 4 \\ 2 \end{pmatrix}$ (b) $\begin{pmatrix} -5 \\ 3 \end{pmatrix}$ (c) $\begin{pmatrix} 0 \\ 4 \end{pmatrix}$ (d) $\begin{pmatrix} -3 \\ -2 \end{pmatrix}$ (e) $\begin{pmatrix} -5 \\ 0 \end{pmatrix}$

2. (a) $3i + 2j$ (b) $i + 3j$ (c) $-2i$ (d) $4i - 5j$ (e) $2j$

3. (a) $\begin{pmatrix} 2 \\ 1 \end{pmatrix}$ (b) $\begin{pmatrix} 2 \\ -3 \end{pmatrix}$ (c) $\begin{pmatrix} -4 \\ 1 \end{pmatrix}$

4. $x = 5, y = 3$ 5. $(6, -2)$ 6. $(1, -7)$ 7. $m = 1, n = 2$

8. $3i + 4j$ 9. $17i - 2j$ 10. $6i + j$

Answers to Exercises

Exercise 13.2

1. 085.2°, 602.1 km h^{-1} 2. 237.5 km h^{-1}, 127.3°   3. 436.2 km h^{-1}, 309.9°

4. 033.7°, 208.3 km h^{-1} 5. 151.5°, 3.46 hours   6. (a) 208.7°   (b) 2.49 hours

7. 53.1°, 50 s   8. 036.9°, 40s   9. 139.1°, 97.5 km   10. 5.1 ms^{-1}, 76.5°

Exercise 13.3(a)

1. 2.24, 63.4°   2. $2i - 8j$ 8.25, 166.0°   3. 19.2 km h^{-1}, 098.7°

4. 943.4 km h^{-1}, 118°   5. 229.9 km h^{-1}, 351.8°   6. 28 km h^{-1} 45.5°

Exercise 13.3(b)

1. 229°, 4.54 minutes   2. 228°, 0.63 hours   3. 287°

4. 321°, 3.84 minutes   5. 56.4°, 11.25 a.m.   6. 9.6°

Review exercise 13

1. (a) $i + 8j$ (b) $7i - 15j$ (c) $13i$

2. $x = -1, y = 3$ 3. $m = 2, n = 3$

4. $p = -1, q = -15$ 5. $m = 3, n = -2$

6. (a) $\sqrt{41}$ (b) $\sqrt{221}$ (c) $\sqrt{65}$ 7. (a) 5   (b) 13   (c) $\sqrt{13}$ (d) $\sqrt{41}$

8. (a) $-\frac{3}{5}i + \frac{4}{5}j$ (b) $\frac{3\sqrt{10}}{10}i - \frac{\sqrt{10}}{10}j$ (c) $\frac{2\sqrt{5}}{5}i + \frac{\sqrt{5}}{5}j$ (d) $-\frac{5}{13}i + \frac{12}{13}j$

9. (a) $\vec{BA} = \begin{pmatrix} -4 \\ 4 \end{pmatrix}$ $\vec{BC} = \begin{pmatrix} 1 \\ 5 \end{pmatrix}$

10. $\vec{AB} = -i + 3j$, $\vec{BC} = -3i + 3j$ $\vec{CA} = 4i - 6j$

11. $-2i - 5j$ 12. $6i + 3j$ 13. $-2i + 17j$

14. $i + 6j$ 15. $m = 1, n = 2$ 16. $3i + 2j$, 5, 36.9°

19. $x = 1, y = -9$ 21. 049°, 227.5 km h^{-1} 22. 338°, 3 hours

23. 9.4 m s^{-1}, 58°   24. (a) 13 m s^{-1} (b) 8.25 s   25. 4.12, 76°

26. 20.7 km h^{-1} 27. 036.9°, 4 hours   28. 291.2°, 2.3 hours

Chapter 14

Exercise 14.1(a)

1. (a) $x = 3, y = 1, z = 2$ (b) $x = 2, y = 3$

2. (a) $\begin{pmatrix} 11 & 7 \\ 2 & 5 \end{pmatrix}$ (b) $\begin{pmatrix} 6 & 1 \\ 9 & 1 \end{pmatrix}$ (c) $\begin{pmatrix} 5 & 5 \\ 1 & 1 \end{pmatrix}$ (d) $\begin{pmatrix} 3 & 5 \\ 1 & 4 \end{pmatrix}$

(e) $\begin{pmatrix} -2 & 1 & -1 \\ 2 & 3 & -3 \end{pmatrix}$ (f) $\begin{pmatrix} 1 & -1 & 1 \\ 2 & 4 & -2 \end{pmatrix}$ (g) $\begin{pmatrix} 1 & -1 \\ 5 & 2 \\ 1 & -1 \end{pmatrix}$

(h) $\begin{pmatrix} 2 & 2 & -4 \\ 2 & 4 & 2 \\ -1 & 1 & -1 \end{pmatrix}$

3. $a = 1, b = -1, c = -5, d = 3$

4. $w = -1, x = -2, y = 5, z = 3$

5. $x = 2, y = 1$ 6. $x = 1, y = 1$

7. (a) $\begin{pmatrix} 2 & -3 \\ -1 & 4 \end{pmatrix}$ (b) $\begin{pmatrix} 3 & 2 \\ -5 & 1 \end{pmatrix}$ (c) $\begin{pmatrix} -6 & 0 \\ 0 & 6 \end{pmatrix}$

(d) $\begin{pmatrix} -3 & 2 & 0 \\ 1 & -5 & 2 \end{pmatrix}$ (e) $\begin{pmatrix} 4 & 0 \\ -3 & -1 \\ 2 & 1 \end{pmatrix}$ (f) $\begin{pmatrix} -2 & 5 & -3 \\ 3 & 0 & 2 \\ -4 & 3 & -2 \end{pmatrix}$

8. (a) $\begin{pmatrix} 1 & 1 \\ 0 & 1 \end{pmatrix}$ (b) $\begin{pmatrix} 3 & -6 \\ -2 & 3 \end{pmatrix}$ (c) $\begin{pmatrix} 7 & -4 \\ -3 & 4 \end{pmatrix}$

(d) $\begin{pmatrix} 0 & 4 \\ -3 & 2 \end{pmatrix}$ (e) $\begin{pmatrix} 1 & 8 & -3 \\ 6 & -5 & 0 \end{pmatrix}$ (f) $\begin{pmatrix} 8 & 4 & 4 \\ 1 & -1 & 0 \end{pmatrix}$

(g) $\begin{pmatrix} 11 & 1 \\ -10 & 1 \\ 3 & -5 \end{pmatrix}$ (h) $\begin{pmatrix} 5 & 1 & -1 \\ -4 & 2 & 2 \\ 7 & -12 & 3 \end{pmatrix}$

9. (a) $\begin{pmatrix} 3 & 6 \\ -6 & 3 \end{pmatrix}$ (b) $\begin{pmatrix} 6 & -2 \\ -2 & 4 \end{pmatrix}$ (c) $\begin{pmatrix} -4 & 0 \\ 0 & 4 \end{pmatrix}$

(d) $\begin{pmatrix} 1 & -3 \\ -2 & 0 \end{pmatrix}$ (e) $\begin{pmatrix} -4 & 0 \\ 8 & -6 \end{pmatrix}$ (f) $\begin{pmatrix} -6 & 3 \\ 9 & -3 \end{pmatrix}$

(g) $\begin{pmatrix} -4 & 2 & 6 \\ 0 & -6 & 4 \end{pmatrix}$ (h) $\begin{pmatrix} 0 & 3 \\ -3 & 6 \\ -6 & 9 \end{pmatrix}$ (i) $\begin{pmatrix} -1 & 0 & 2 \\ 1 & -3 & 1 \\ 0 & 1 & -2 \end{pmatrix}$

10. (a) $\begin{pmatrix} -8 & 5 \\ 5 & 5 \end{pmatrix}$ (b) $\begin{pmatrix} 4 & -11 \\ 1 & -4 \end{pmatrix}$ (c) $\begin{pmatrix} 9 & -12 \\ -4 & 4 \end{pmatrix}$

(d) $\begin{pmatrix} 2 & -9 \\ -1 & 12 \end{pmatrix}$ (e) $\begin{pmatrix} 0 & -4 \\ 0 & 17 \end{pmatrix}$

11. $\begin{pmatrix} 25 & 18 & 12 \\ 30 & 45 & 28 \end{pmatrix}$ 12. $\begin{pmatrix} 315 & 585 \\ 420 & 280 \\ 184 & 216 \end{pmatrix}$

13. $\begin{pmatrix} 285 & 175 \\ 276 & 143 \end{pmatrix}$

Shop A sold 285 shirts and 175 pair of trousers

Shop B sold 276 shirts and 143 pair of trousers

14. $\begin{pmatrix} 30 & 15 \\ 18 & 16 \end{pmatrix}$

Law faculty: 30 males, 15 female, Science faculty: 18 males, 16 females

Exercise 14.1(b)

1. (a) $\begin{pmatrix} 11 & 0 \\ 12 & -3 \end{pmatrix}$ (b) $\begin{pmatrix} -7 & -2 \\ -2 & -8 \end{pmatrix}$ (c) $\begin{pmatrix} 6 & -14 \\ 10 & -16 \end{pmatrix}$

(d) $\begin{pmatrix} 38 & 6 \\ 5 & 0 \end{pmatrix}$ (e) $\begin{pmatrix} 0 & 18 \\ 0 & -12 \end{pmatrix}$ (f) $\begin{pmatrix} 0 & 0 \\ 0 & 0 \end{pmatrix}$

(g) $\begin{pmatrix} 8 \\ 16 \end{pmatrix}$ (h) $\begin{pmatrix} -1 \\ -1 \end{pmatrix}$ (i) $\begin{pmatrix} -14 \\ 3 \end{pmatrix}$

(j) $\begin{pmatrix} -1 \\ 3 \end{pmatrix}$ (k) $\begin{pmatrix} -10 & -4 \\ -7 & -4 \end{pmatrix}$ (l) $\begin{pmatrix} 8 & -4 \\ 8 & -2 \\ -3 & 1 \end{pmatrix}$

(m) $\begin{pmatrix} 1 & -7 \\ 4 & -2 \\ 2 & -8 \end{pmatrix}$ (n) $\begin{pmatrix} 11 & -7 & 5 \\ 11 & 0 & 1 \\ 6 & 3 & 0 \end{pmatrix}$

2. (a) $\begin{pmatrix} -2 & -3 \\ 9 & 1 \end{pmatrix}, \begin{pmatrix} -11 & -4 \\ 12 & -7 \end{pmatrix}$ (b) $\begin{pmatrix} 7 & 2 \\ -4 & -1 \end{pmatrix}, \begin{pmatrix} -17 & -5 \\ 10 & 3 \end{pmatrix}$

(c) $\begin{pmatrix} 1 & 0 \\ 0 & 1 \end{pmatrix}, \begin{pmatrix} 1 & 0 \\ 2 & -1 \end{pmatrix}$

3. $x = 2, y = -5, z = -11$ 4. $x = 3, y = -2, z = 7$

5. $a = -2, b = 1, c = -1, d = 2$ 6. $a = 2, b = -3, c = 3, d = -2$

7. Shop A makes £1800 profit, Shop B makes £1988

8. John spent £170, Eric spent £148.50

Exercise 14.1(c)

1. (a) −1   (b) 2   (c) 2   (d) −1   (e) 0   (f) −2

2. (a) −7   (b) −14   (c) 98   (d) 98   (e) 98

3. (a) −5   (b) −1, 1   (c) 1, 6

4. (a) $\begin{pmatrix} 0 & 1/2 \\ 1/3 & 1/3 \end{pmatrix}$ (b) $\begin{pmatrix} 3/2 & -5/2 \\ -1 & 2 \end{pmatrix}$

   (c) $\begin{pmatrix} -3/2 & -1/2 \\ 2 & -1 \end{pmatrix}$ (d) $\begin{pmatrix} -5/2 & 2 \\ 3/2 & -1 \end{pmatrix}$

   (e) $\begin{pmatrix} 1/5 & 1/5 \\ 0 & 1/2 \end{pmatrix}$ (f) $\begin{pmatrix} 1/4 & -1/4 \\ 1/8 & 3/8 \end{pmatrix}$

5. (a) $\begin{pmatrix} 1/5 & 2/5 \\ 2/5 & -1/5 \end{pmatrix}$ (b) $\begin{pmatrix} -1/11 & 4/11 \\ 3/11 & -1/11 \end{pmatrix}$

   (c) $\begin{pmatrix} 1 & 0 \\ 0 & 1 \end{pmatrix}$ (d) $\begin{pmatrix} 1 & 0 \\ 0 & 1 \end{pmatrix}$

6. (a) $\begin{pmatrix} -1 & 2 \\ 3/2 & -5/2 \end{pmatrix}$ (b) $\begin{pmatrix} -2 & 6 \\ 1 & -13 \end{pmatrix}$

   (c) $\begin{pmatrix} 1/2 & 1 \\ -3/4 & -2 \end{pmatrix}$ (d) $\begin{pmatrix} 1/4 & 4 \\ -39/4 & -7 \end{pmatrix}$

Exercise 14.2

1. (a) $x = -1, y = 2$ (b) $x = 1, y = 0$ (c) $x = 2, y = 3$

   (d) $x = -2, y = 2$ (e) $x = 3, y = -2$ (f) $x = 2, y = 3$

2. 12 of the $0.10 coins, 10 of the $0.05 coins

3. 300 of the £5 tickets, 200 of the $8 tickets

4. £500 was invested in bonds, £300 was invested in a saving account

5. 200 ml of 15% acid solution, 300 ml of 25% acid solution.

6. 120 km h^{-1}, 60 km h^{-1}

Review exercise 14

1. (a) $\begin{pmatrix} 4 & 1 \\ 4 & 6 \end{pmatrix}$ (b) $\begin{pmatrix} 5 & 2 \\ 8 & 0 \end{pmatrix}$ (c) $\begin{pmatrix} -4 & -5 \\ -12 & 14 \end{pmatrix}$

(d) $\begin{pmatrix} 3 & 5 \\ 4 & 2 \end{pmatrix}$ (e) $\begin{pmatrix} -3 & -4 \\ -8 & 8 \end{pmatrix}$ (f) $\begin{pmatrix} -6 & 13 \\ -4 & -14 \end{pmatrix}$

(g) $\begin{pmatrix} 0 & -5 \\ -4 & 10 \end{pmatrix}$ (h) $\begin{pmatrix} 6 & 15 \\ 12 & -6 \end{pmatrix}$ (i) $\begin{pmatrix} -23 & -6 \\ -40 & 8 \end{pmatrix}$

(j) $\begin{pmatrix} -6 & 3 \\ -12 & 6 \end{pmatrix}$ (k) $\begin{pmatrix} 1 & -10 \\ 0 & 4 \end{pmatrix}$ (l) $\begin{pmatrix} 3 & -8 \\ 8 & -8 \end{pmatrix}$

2. (a) $\begin{pmatrix} 12 & 10 \\ 10 & -8 \end{pmatrix}$ (b) $\begin{pmatrix} 4 \\ 2 \end{pmatrix}$ (c) $\begin{pmatrix} 20 & -8 \\ 2 & 2 \end{pmatrix}$

(d) $\begin{pmatrix} -4 \\ 7 \end{pmatrix}$ (e) $\begin{pmatrix} 16 & -4 \\ 18 & -10 \end{pmatrix}$ (f) $\begin{pmatrix} 12 \\ 1 \end{pmatrix}$

(g) $\begin{pmatrix} 22 & 12 \\ 12 & 4 \end{pmatrix}$ (h) $\begin{pmatrix} -2 & 14 \\ -4 & 0 \end{pmatrix}$ (i) $\begin{pmatrix} 20 \\ -4 \end{pmatrix}$

3. (a) $\begin{pmatrix} 9 & -1 \\ 0 & 4 \end{pmatrix}$ (b) $\begin{pmatrix} -27 & 7 \\ 0 & 8 \end{pmatrix}$

4. (a) $\begin{pmatrix} 10 & -6 \\ 9 & -5 \end{pmatrix}$ (b) $\begin{pmatrix} -3 & 4 \\ -4 & -3 \end{pmatrix}$ (c) $\begin{pmatrix} 22 & 14 \\ 19 & 13 \end{pmatrix}$ (d) $\begin{pmatrix} 0 & 2 \\ -25 & 11 \end{pmatrix}$

5. $a = 3, b = 1, c = 1, d = -2$ 6. $x = -5, y = 9$ 7. $x = 4, y = 5$

8. (a) 2   (b) 7   (c) 28   (d) 14   (e) 2   (f) – 6

9. (a) – 5   (b) – 1, 1   (c) 1, 6   (d) –3/2, 4

10. (a) $\begin{pmatrix} 1 & -1 \\ -1 & 2 \end{pmatrix}$ (b) $\begin{pmatrix} 1/5 & 1/5 \\ -2/5 & 3/5 \end{pmatrix}$ (c) $\begin{pmatrix} 1 & -5/2 \\ -1 & 3 \end{pmatrix}$

(d) $\begin{pmatrix} -1 & -1/2 \\ -3 & -2 \end{pmatrix}$ (e) $\begin{pmatrix} 2 & 1 \\ -3/2 & -1/2 \end{pmatrix}$ (f) $\begin{pmatrix} -3 & 2 \\ 2 & -1 \end{pmatrix}$

11. (a) $\begin{pmatrix} 0 & 2 \\ -2 & 4 \end{pmatrix}$ (b) $\begin{pmatrix} 5 & -2 \\ 2 & 1 \end{pmatrix}$ 12. $\begin{pmatrix} 0.18 & -0.14 \\ -0.07 & 0.11 \end{pmatrix}$

13. (a) $\begin{pmatrix} 1/5 & 2/5 \\ 2/15 & -1/15 \end{pmatrix}$ (b) $\begin{pmatrix} 1/31 & -4/31 \\ 6/31 & 7/31 \end{pmatrix}$

14. $\begin{pmatrix} -2 & 1 \\ 3/2 & -1/2 \end{pmatrix}$ 15. $\begin{pmatrix} 3/7 & -1/7 \\ 1/7 & 2/7 \end{pmatrix}, \begin{pmatrix} 3 & 5 \\ -5 & 8 \end{pmatrix}$

16. (a) $x = 3, y = -4$ (b) $x = 3, y = 2$

  (c) $x = -1, y = 2$ (d) $x = 2, y = 1$

17. $\begin{pmatrix} 1/3 & 2/3 \\ 1/2 & 1/2 \end{pmatrix}$ (a) $x = 3, y = 2$ (b) $x = -5/3, y = 1$

18. $\begin{pmatrix} 2/17 & 3/17 \\ 3/17 & -4/17 \end{pmatrix}$ (a) $x = -2, y = 3$ (b) $x = 2, y = 1$

Chapter 15

Exercise 15.1(a)

1. (a) 1   2. 5   3. $6x$ 4. $-4x$ 5. 0   6. $-x$

7. $2x^2$ 8. $-6x$ 9. $-6x^2$

Exercise 15.1(b)

1. -1 2. 0   3. $4x^3$ 4. $6x^5$ 5. $3x^{-4}$

6. $-x^{-2}$ 7. $-8x^{-9}$ 8. $-\frac{3}{4}x^{-7/4}$ 9. $-\frac{1}{2}x^{-1/2}$

10. $-\frac{3}{2}x^{-5/2}$ 11. $\frac{2}{3}x^{-5/3}$ 12. $-\frac{1}{3}x^{-2/3}$

Exercise 15.1(c)

1. -2 2. $12x^3$ 3. $-10x^4$ 4. $2x^7$

5. $4x^5$ 6. $2x^{-6}$ 7. $-6x^{-4}$ 8. $12x^{-4}$

9. $-6x^{-3}$ 10. $x^{-2/3}$ 11. $-3x^{1/2}$ 12. $3x^{-1/2}$

13. $4x^{-5/2}$ 14. $-6x^{-7/3}$ 15. $-6x^{-5/3}$

Exercise 15.1(d)

1. $6x^2 + 12x^3$ 2. $15x^4 - 2$ 3. $-4t^3 + 5t^4$

4. $2t^{-2}$ 5. $12t^2 + 3$ 6. $20x^3 - 15x^4$ 7. $3x^{-4}$

8. $6x^2 + 6x^{-3}$ 9. $12t^3 - 4$ 10. $4t - 3$ 11. $12x + 5$

12. $t^2 + t^{-3}$ 13. $x^{-2/3} + x^{-3/2}$ 14. $-3t^{-7/4} + t^2$

340 Answers to Exercises

15. $4x^{-5/3}$ 16. $-3x^{-4}$ 17. $3 + \frac{1}{2}x^{-3/2} - x^{-2}$

18. $-10x^{-3} + \frac{3}{2}x^{-5/2}$ 19. $2t - 12t^{-4}$ 20. $6 - 4x$

21. $6x - 12$ 22. $1 - \frac{4}{3}x$ 23. $3 + x^{-2}$

24. $2 + 5t^{-1} - 3t^{-2}$

Exercise 15.1(e)

1. $20x^4 - 9x^2$ 2. $4x + 5$ 3. $18x^2 + 6x$

4. $12x - 1$ 5. $8x^3 - 9x^2$ 6. $6 - 2x - 6x^2$

7. $3x^{1/2} + \frac{3}{2}x^{-1/2}$ 8. $2x - 8x^3$ 9. $\frac{2}{3}x^{-2/3} + \frac{4}{3}x^{1/3}$

10. $-6x^{-2} + 18x^{-3}$ 11. $7x^6 + 4x^3 - 3x^2$

12. $18x^2 - 30x + 4$

Exercise 15.1(f)

1. $\dfrac{1}{(x+1)^2}$ 2. $\dfrac{-2}{(x-1)^2}$ 3. $\dfrac{-2x}{(x^2-1)^2}$

4. $\dfrac{-12x}{(3x^2-1)^2}$ 5. $\dfrac{-3x^2-6}{(x^2-2)^2}$ 6. $\dfrac{6x^2-42x+10}{(2x-7)^2}$

7. $\dfrac{-1}{x^{1/2}(1+x^{1/2})^2}$ 8. $\dfrac{4x}{(1-x^2)^2}$ 9. $\dfrac{x^2+2x}{(x+1)^2}$

10. $\dfrac{-2x}{(x^2-1)^2}$ 11. $\dfrac{2x^2+6x}{(2x+3)^2}$ 12. $\dfrac{-1}{2x^{1/2}(x^{1/2}-1)^2}$

Exercise 15.1(g)

1. $3(x-5)^2$ 2. $15(3x-2)^4$ 3. $-24x(1-2x^2)^5$

4. $8x(x^2+1)^3$ 5. $21x^2(x^3-2)^6$ 6. $-6(2x+1)^{-4}$

7. $8(3-2x)^{-5}$ 8. $-8(4x+3)^{-3}$ 9. $4x(1-2x^2)^{-2}$

10. $x(x^2-1)^{-1/2}$ 11. $-2(3x+2)^{-5/3}$

12. $3(2x^2-3)^{-1/4}$ 13. $\dfrac{-4(3x+1)}{(3x^2+2x)^3}$ 14. $\dfrac{1}{3(1-x)^{4/3}}$

Answers to Exercises 341

15. $\dfrac{-x}{(x^2-1)^{3/2}}$ 16. $\dfrac{-3x^2}{2(2+x^3)^{3/2}}$

Exercise 15.1(h)

1. $4(x+3)^3$ 2. $9(3x-2)^2$ 3. $20x(2x^2+3)^4$

4. $x(x^2+1)^{-1/2}$ 5. $-2x^2(2x^3-1)^{-4/3}$

6. $-4x(3+x^2)^{-3}$ 7. $-6x(1-x^2)^2$

8. $-18x(3x^2-1)^{-4}$ 9. $-12x(1-x^2)^5$

10. $-8x(x^2-3)^{-5}$ 11. $-2x(2x^2+3)^{-3/2}$

12. $-(x^2+1)(x^3+3x)^{-4/3}$

Exercise 15.1(i)

1. (a) $-2\sin 2x$ (b) $5\cos 5x$ (c) $3\sec^2 3x$

   (d) $-6\sin 3x$ (e) $-3\cos 6x$ (f) $-8\operatorname{cosec}^2 2x$

   (g) $-3\sin(3x-2)$ (h) $2\operatorname{cosec}^2(5-2x)$

   (i) $2\cos(2x+3)$ (j) $2x\cos x^2$ (k) $-3\sin\tfrac{1}{2}x$

   (l) $6x\sec^2 3x^2$ (m) $2\sec^2(2x-5)$

   (n) $-\tfrac{2}{3}x\operatorname{cosec}^2\tfrac{1}{3}x^2$ (o) $2x\sec^2(x^2+1)$

2. (a) $-4\cos 2x\sin 2x$ (b) $6\sin x\cos x$

   (c) $4\tan 2x\sec^2 2x$ (d) $6\sin 3x\cos 3x$

3. (a) $-x\sin x+\cos x$ (b) $2x\cos 2x+\sin 2x$

   (c) $x^2\cos x+2x\sin x$ (d) $x^3\sec^2 x+3x^2\tan x$

4. (a) $x\cos\tfrac{1}{2}x+2\sin\tfrac{1}{2}x$ (b) $-2x\sin 2x+\cos 2x$

   (c) $-x^2\sin x+2x\cos x$ (d) $2x^2\sec^2 2x+2x\tan 2x$

   (e) $-\sin^2 x+\cos^2 x$ (f) $-2x^2\operatorname{cosec}^2 2x+2x\cot 2x$

   (g) $\cos x\sec^2 x-\tan x\sin x$

(h) $\dfrac{x\cos x - \sin x}{x^2}$ (i) $\dfrac{-2x\sin 2x - \cos 2x}{x^2}$

5. (a) $2\cos x + 3$ (b) $2\sin 2x$ (c) $2x + 3\sec^2 x$

(d) $-2\sin x$ (e) $2\cos x + 2x$ (f) $6x^2 + 3\sin x$

(g) $4x + 5\sec^2 x$ (h) $6\cos 2x + \sin x$

(i) $x\sec^2 x + \tan x - 6x$ (j) $x^2\cos x + 2x\sin x$

(k) $x^2\sec^2 x + 2x\tan x - \dfrac{1}{x^2}$ (l) $x^2\cos x + 2x\sin x + \dfrac{2}{x^3}$

Exercise 15.1(j)

1. $\dfrac{1}{x}$ 2. $\dfrac{1}{x}$ 3. $\dfrac{2}{x}$ 4. $\dfrac{3}{3x+2}$ 5. $\dfrac{-2}{1-2x}$

6. $\dfrac{1}{4x+3}$ 7. $\dfrac{3}{x}$ 8. $\dfrac{1}{3x}$ 9. $\dfrac{-1}{2x}$ 10. $\dfrac{-3}{x}$

11. $\dfrac{-2}{x}$ 12. $\dfrac{-2}{3x}$ 13. $2 + \ln x^2$ 14. $x^2 + 3x^2\ln x$

15. $\dfrac{2x}{2x-1} + \ln(2x-1)$ 16. $x + 2x\ln\tfrac{1}{2}x$ 17. $2 + \ln 3x^2$

18. $\dfrac{3x^3}{3x+2} + 3x^2\ln(3x+2)$

Exercise 15.1(k)

1. $3e^{3x}$ 2. $\tfrac{1}{2}e^{\tfrac{1}{2}x}$ 3. $-\tfrac{3}{4}e^{-\tfrac{3}{4}x}$ 4. $-2e^{-2x}$

5. $3e^{3x+2}$ 6. $-2e^{1-2x}$ 7. $\tfrac{1}{3}e^{\tfrac{1}{3}x+2}$ 8. $-\tfrac{2}{3}e^{-\tfrac{2}{3}x+5}$

9. $6xe^{3x^2}$ 10. $-3x^2e^{1-x^3}$ 11. $\dfrac{1}{2\sqrt{x}}e^{\sqrt{x}}$ 12. $\dfrac{2}{3\sqrt[3]{x}}e^{\sqrt[3]{x^2}}$

13. $4x^2e^{2x^2} + e^{2x^2}$ 14. $\tfrac{1}{2}\sqrt{x^3}e^{\sqrt{x}} + 2xe^{\sqrt{x}}$

15. $-\tfrac{2}{3}x^2e^{-\tfrac{1}{3}x^2} + e^{-\tfrac{1}{3}x^2}$ 16. $xe^x + 2e^x$

17. $-2x^2e^{1-x^2} + e^{1-x^2}$ 18. $3x^4e^{x^3} + 2xe^{x^3} - 3x^2e^{x^3}$

Exercise 15.1(l)

1. (a) 3 (b) –8 (c) 0 (d) 9 (e) 4

(f) –9 (g) 0 (h) –5/2

2. (a) $y = -4x + 1$ (b) $y = 9x - 23$ (c) $y = -4x + 9$

(d) $y = x$ (e) $y = x + 7$ (f) $y = x$

3. (a) $x + 5y - 21 = 0$ (b) $x - 8y + 42 = 0$

(c) $y = 6$ (d) $x + 5y - 16 = 0$

(e) $x + 23y + 71 = 0$ (f) $x + 12y - 13 = 0$

4. $(1/2, -9/4)$ 5. $(2, -3)$

6. $(-2/3, -32/27), (2, 0)$ 7. $(2/3, 26/9)$

Exercise 15.1(m)

1. (a) x/y (b) $1/y$ (c) $\frac{2y-x}{y-2x}$ (d) $\frac{x+y}{y-x}$ (e) x^2/y^2

(f) $\frac{2x-2y}{2x}$ (g) $\frac{x+4}{y-3}$ (h) $\frac{2x+3y}{2y-3x}$ (i) $\frac{2x+2y+2}{3-2y-2x}$ (j) $-3y/2x$

2. (a) $-2/3$ (b) $1/3$ (c) $-1/4$ (d) -9 (e) $-7/4$ (f) -1

3. (a) $y = -4x + 14$, $x - 4y + 5 = 0$

(b) $8x + 7y - 22 = 0$, $7x - 8y + 9 = 0$

(c) $x + 8y - 26 = 0$, $y = 8x - 13$

(d) $y = -4x + 7$, $x - 4y - 6 = 0$

Exercise 15.1(n)

1. 6 2. 0 3. $6x - 6$ 4. $12x^2 + 2$ 5. 6 6. $2 - 6x$

7. $20x^{-6}$ 8. $24x^{-5}$ 9. $6x^{-3}$ 10. $12x^{-5}$ 11. $12x^{-4}$

12. $-48x^{-5}$ 13. $12x + 30x^4$ 14. $-1/2\sqrt{x^3}$ 15. $24(2x - 3)$

16. $-2/(x + 1)^3$ 17. $\frac{4}{(1-x)^3}$ 18. $\frac{4(1+3x^2)}{(1-x^2)^3}$

Answers to Exercises

Exercise 15.1(o)

1. $(2, -1)$ min. 2. $(1, 6)$ max. 3. $(0, 0)$ min. $(4/3, 32/27)$ max.

4. $(0, 0)$ min. $(2, 4)$ max. 5. $(1, -9)$ min. $(-1, -5)$ max.

6. $(3, -27)$ min. $(0, 0)$ max. 7. $(-2, 26)$ max. $(1, -1)$ min.

8. $(-1, 8)$ max. $(4/3, -127/27)$ min. 9. $(-5, -175/3)$ min. $(3, 27)$ max.

10. $(2, 14/3)$ max. $(3, 9/2)$ min. 11. $(0, 0)$ max. $(4/3, -32/27)$ min.

12. $(2/3, 100/27)$ max. $(3, -9)$ min.

Exercise 15.1(p)

1. $(3, -9)$ min. 2. $(1/8, 1/16)$ max. 3. $(2, 3)$ min. 4. $(3, 14)$ max.

5. $(-1, 8)$ max. $(4/3, -67/27)$ min. 6. $(1, 2)$ min. $(-1, -2)$ max.

7. $(-1/3, 62/27)$ max. $(4, -101)$ min. 8. $(-1, -5)$ min.

9. $(-4/5, -16/5)$ min. 10. $(1, 5)$ min. $(-1, 9)$ max.

11. $(1, 12)$ max. $(3, 9)$ min. 12. $(1, 2)$ max. $(-1, -2)$ min.

13. $(0, 4)$ max. $(2, 0)$ min. 14. $(-2/3, -16/9)$ min. $(2/3, 16/9)$ max.

15. $(0, 0)$ min. $(4/3, 32/27)$ max. 16. $(1/2, 4)$ min. $(-1/2, -4)$ max.

17. $(1/2, 3/2)$ min. 18. $(-2, -3)$ max.

Exercise 15.1(q)

1. 8 cm   2. 9.85   3. 20   4. 2.17   5. Radius = 6.83 cm; height = 93.24 cm

6. 1.47 cm   7. 81/4   8. 8 cm by 8 cm   9. 5000 m^2, 50 cm by 100 cm

10. 1200 cm, 30 cm by 60 cm   11. 1 cm   12. 2 cm   13. 2 cm

14. £ 3600   20 cm by 20 cm by 6 cm

15. $ 600   16. 15 cm by 15 cm by 30 cm, £4050

17. 1.37 cm, 23. 43 cm, £ 70.30

Exercise 15.1(r)

1.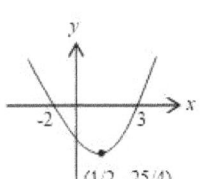
2.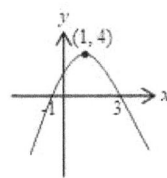
3.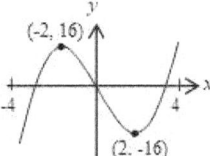
4.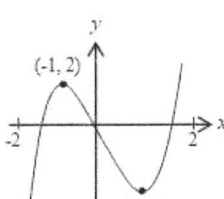
5.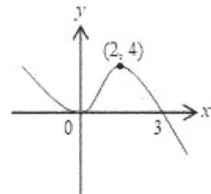
6.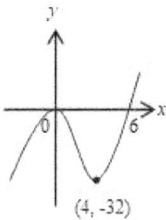
7.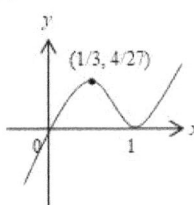
8.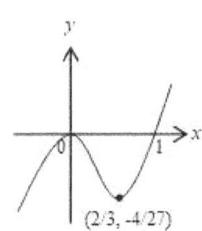
9.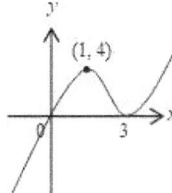
10.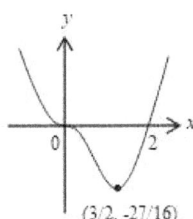
11.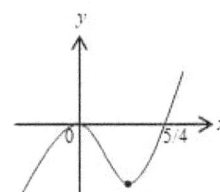
12.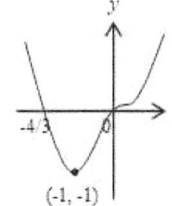

Exercise 15.1(s)

1. 2 cm 2. 3.84 cm^2 3. 5.07π cm^3 4. 6 %

5. 2.5 % 6. 2.04 m 7. 3 % 8. 4 %, 6 %

9. 0.72π cm 10. 4/3 % 11. 2,325 12. 3.011

Answers to Exercises

13. 2.000125   14. 4.15   15. 3.00185

16. 3.074   17. 12.125   18. 2.05   19. 10.0233

Exercise 15.1(t)

1. $16 \text{ cm}^2 \text{ s}^{-1}$ 2. $4.8\pi \text{ cm}^3 \text{ s}^{-1}, 7.2\pi \text{ cm}^3 \text{ s}^{-1}$

3. $3/10\pi \text{ cm s}^{-1}$ 4. $2/45\pi \text{ cm s}^{-1}$ 5. $1/8 \text{ cm s}^{-1}, 6 \text{ cm}^2 \text{ s}^{-1}$

6. $1000\pi \text{ cm}^3 \text{ s}^{-1}$ 7. $1/200\pi \text{ m s}^{-1}$ 8. $1/2\pi \text{ cm s}^{-1}$

9. $1/20\pi \text{ cm s}^{-1}$ 10. $1/5\pi \text{ cm s}^{-1}$

Exercise 15.2(a)

1. $5x + c$ 2. $x/3 + c$ 3. $x^3/3 + c$ 4. $x^5/5 + c$

5. $x^{10}/10 + c$ 6. $x^{12}/12 + c$ 7. $-x^{-2}/2 + c$

8. $-x^{-1} + c$ 9. $-x^{-4}/4 + c$ 10. $3x^{4/3}/4 + c$

11. $4x^{7/4}/7 + c$ 12. $2x^{5/2}/5 + c$ 13. $4x^{3/4}/3 + c$

14. $3x^{1/3} + c$ 15. $-3x^{-1/3} + c$ 16. $5x^{6/5}/6 + c$

17. $3x^{5/3}/5 + c$ 18. $4x^{9/4}/9 + c$ 19. $-x^{-3}/3 + c$

20. $-x^{-6}/6 + c$ 21. $-x^{-5}/5 + c$ 22. $5x^{1/5} + c$

23. $-2x^{-1/2} + c$ 24. $4x^{1/4} + c$

Exercise 15.2(b)

1. $3x^2 + c$ 2. $2x^6/3 + c$ 3. $2x^4 + c$

4. $x^3/4 + c$ 5. $x^5/4 + c$ 6. $x^8/12 + c$

7. $-4x^{-2} + c$ 8. $-4x^{-1} + c$ 9. $x^3 + c$

10. $-x^{-1}/3 + c$ 11. $-2x^{-3}/9 + c$ 12. $6x^{1/2} + c$

Exercise 15.2(c)

1. $x^2 - 3x + c$ 2. $3x^2/2 + 5x + c$ 3. $x^3/3 - 8x + c$

4. $x^5/5 + x^4/4 + c$ 5. $2x^3/3 + x^4/4 + c$

6. $x^4/2 - x^5 + c$ 7. $x^3 - x^2 + x + c$

8. $3x - x^2 + 4x^3/3 + c$ 9. $x - x^2 + 2x^3 + c$

10. $x^5 - x^3 + x^2 + c$ 11. $x^4/4 - x^2 + c$

12. $x^4 + x^3 + c$ 13. $2x^3/3 + 3x^5/5 + c$

14. $x^6/6 - x^3 + c$ 15. $x^4 - 2x^5/5 + c$

16. $3x^5/5 + x^3 + c$ 17. $x^3/3 - x^2 - 3x + c$

18. $x^3 + 7x^2/2 + 3x + c$ 19. $9x - 6x^2 + 4x^3/3 + c$

20. $x^4/4 - 4x^3/3 + 2x^2 + c$ 21. $2x^{7/2}/7 - 2x^{5/2}/5 + c$

22. $3x^{10/3}/10 + 6x^{7/3}/7 + c$ 23. $x^2/2 - 1/x + c$

24. $5x - 3x^2/2 + c$ 25. $-1/2x^2 - 2/3x^3 + c$

26. $3x^{8/3}/8 - 12x^{5/3}/5 + 3x^{2/3}/2 + c$

27. $-2/x^{1/2} + 5/x + c$ 28. $2x^{1/2} + 2x^{5/2}/5 + c$

29. $-4/x - 3x + x^2 + c$ 30. $-1/x - x - 2x^3/3 + c$

Exercise 15.2(d)

1. 3/2   2. 26/3   3. 60   4. 27   5. 31/5

6. 1   7. 16/3   8. 16/3   9. 44   10. 9/4

11. 6   12. −4/3   13. 65/6   14. 87/2

15. 318/7   16. 10   17. 2   18. 26/15

19. 3/2   20. 4   21. 1/3   22. 3/8   23. 1/9

24. 14/3   25. 14/3   26. 2   27. −3

28. 52/5   29. −2   30. 64/3

Exercise 15.2(e)

1. $(3x + 2)^4/12 + c$ 2. $-1/(2x - 1) + c$

3. $3(x-2)^{4/3}/4 + c$ 4. $(x^2+3)^3/6 + c$

5. $(x^2+1)^5/10 + c$ 6. $(2x^3+3)^4/30 + c$

7. $-(1-x^2)^{3/2}/3 + c$ 8. $2(x^3-2)^{3/2}/9 + c$

9. $(4x^2-5)^{3/2}/12 + c$ 10. $(x^3+4)^{3/2}/9 + c$

11. $-(2-x^4)^{3/2}/6 + c$ 12. $(2x^3-3)^{3/2}/9 + c$

13. $-(2x-5)^{-3}/6 + c$ 14. $(3x+1)^{2/3}/2 + c$

15. $2(1+3x^2)^{1/2}/3 + c$ 16. $-(x^2-1)^{-3}/2 + c$

17. $1/5$ 18. $38/3$ 19. $52/9$ 20. $-3/5$

21. $5/32$ 22. $3/32$

Exercise 15.2(f)

1. $\sin 2x/2 + c$ 2. $-\cos 3x/3 + c$

3. $\tan 4x/2 + c$ 4. $-3\cos 2x/2 + c$

5. $\sin 6x/2 + c$ 6. $3\cos 4x/2 + c$

7. $-\cos 3x/4 + c$ 8. $2\tan(x/2) + c$

9. $\sin(2x+1)/2 + c$ 10. $3\cos(3-2x)/2 + c$

11. $\sin(5x+2)/5 + c$ 12. $\tan(3x-2)/3 + c$

13. $2\cos(3-x/2) + c$ 14. $4\tan(3x/2+2) + c$

15. $-\sin(2-3x)/3 + c$ 16. $-\cos x^3/3 + c$

17. $3\sin x^2/2 + c$ 18. $2\tan x^3/3 + c$

19. $\sin^3 x/3 + c$ 20. $-\cos^4 x/4 + c$

21. 1 22. $\sqrt{2}/2$ 23. $\sqrt{3}$ 24. $-\sqrt{3}/2$

Answers to Exercises

Exercise 15.2(g)

1. $e^{3x}/3 + c$ 2. $-e^{-2x}/2 + c$ 3. $-4e^{-3/4}/3 + c$

4. $3e^{2x/3}/2 + c$ 5. $e^{2x+3}/2 + c$ 6. $-e^{-3x+2}/3 + c$

7. $-2e^{3-4x} + c$ 8. $2e^{x/2+3} + c$ 9. $-4e^{4-x/2} + c$

10. $3e^{4x}/4 + c$ 11. $e^{-4x}/2 + c$ 12. $-3e^{-2x} + c$

13. $e^{x^2}/2 + c$ 14. $2e^{x^3}/3 + c$ 15. $-e^{3-x^2}/2 + c$

Exercise 15.3(a)

1. (a) 75/2 (b) 26/3 (c) 15/12 (d) 12 (e) 8/3 (f) 7/8

2. (a) 19/3 (b) 4 (c) 25/4 3. 125/6 4. 4/3 5. 4/3

6. 1/6 7. 4, 4 8. 8/3 9. 12

10. (a) 9 (b) 12 (c) 7/3 (d) 60

Exercise 15.3(b)

1. (a) 375/6 (b) 36 (c) 367/6 (d) 4/3

(e) 1/6 (f) 125/6 (g) 125/6 (h) 32/3

2. (a) 9 (b) 9 (c) 1/6 (d) 5/12

Exercise 15.4(a)

1. 31 m 36 m s^{-1} 30 m s^{-2}

2. 14 m 15 m s^{-1} 12 m s^{-2} 3. 11 m 32 m

4. 24 m s^{-1} 42 m s^{-2} 5. 19.6 m 6. 3 s $-9/4$

7. 0 s 6 m s^{-1} -10 m s^{-2}; 2 s -2 m s^{-1}

2 m s^{-2}; 3 s 3 m s^{-1} 8 m s^{-2}

8. 3 m s^{-1} -2 m s^{-2} 32/3 m

9(a) 0 m s^{-1} 30 m s^{-2} (b) 31.25 m

10. 27 m -5 m 11. 7.30 a.m. 9 a.m.

Exercise 15.4(b)

1. $s = t^3 + 2t^2 + 2$

2. $v = 3t^3 - 8t + 6$, $s = t^3 - 4t^2 + 6t + 4$

3. 90 m   4. 42 m s^{-1} 30 m   5. 29 m s^{-1} 33 m

6. 192 m   7. 32 m s^{-2} 156 m   8. 144 m   9. 4 s

10. 102.4 km   11. 39.6 m s^{-1} 59.6 m   12. 31.25 m

Exercise 15.5

1. 40 m s^{-1} 2. 5 m s^{-2} 3. 30 s; 1.875 m s^{-2}

4(a) 5 hours   (b) 12 m s^{-2} (c) 18 m s^{-2}

5(a) 7.5 km   (b) 45 km h^{-1} (c) 0.09 m s^{-2}

6(a) 80 s   (b) 0.75 m s^{-2}

Review exercise 15

1. (a) $6x + 2$ (b) $2 - 8x$ (c) $4x^3 + 1$ (d) $4x^3 - 2x$

   (e) $-1/x^{3/2} + 3/x^2$ (f) $10x^4 + 16/x^3 - 2x^{2/3}/3$

2. (a) $2x - 2$ (b) $-8x + 9x^2 - 10x^4$ (c) $-24x + 16x^3$

   (d) $3 - 1/x$ (e) $1/2x^{1/2} + 4x^{1/2}$

   (f) $-x^{5/6}/6 - -x^{3/4}/4 + 3x^{1/2}/2$

3. (a) $3(3x - 7)^{-1/2}/2$ (b) $10x(2 + x^2)^4$

   (c) $-2x(1 - 2x^2)^{-1/2}$ (d) $(1 - 3x)/(1 - 2x)^{1/2}$

   (e) $(4x - 1)/(x^2 - 1)^{3/2}$ (f) $(2x - 1)/2x^{1/2}(x - 1)^{3/2}$

4. (a) $2\cos 2x$ (b) $-5\sin 5x$ (c) $6\cos 3x$

   (d) $3\sin 6x$ (e) $3\cos(3x - 2)$ (f) $3\cos\frac{1}{2}x$

   (g) $-2\sin(2x + 3)$ (h) $3\sec^2(3x + 5)$ (i) $-2x\sin x^2$

   (j) $2x\sec^2(x^2 - 1)$ (k) $2x\cos(x^2 + 3)$ (l) $6x\sec^2 3x^2$

Answers to Exercises   351

5. (a) $x \cos x + \sin x$ (b) $-2x \sin 2x + \cos 2x$

   (c) $2x \sec^2 x + 2 \tan x$ (d) $x^2 \cos x + 2x \sin x$

   (e) $3x^2 \sec^2 3x + 2x \tan 3x$ (f) $-x^3 \sin x + 3x^2 \cos x$

   (g) $(-x \sin x - \cos x)/x^2$ (h) $(2x \cos 2x - \sin 2x)/x^2$

   (i) $-\sin x \sin 2x + 2 \cos x \cos 2x$ (j) $3 \cos x \sec^2 3x - \sin x \tan 3x$

6. (a) $1/x$ (b) $1/x$ (c) $3/x$ (d) $2/(2x - 3)$ (e) $-2e$ (f) $-3x$

7. (a) $-3e^{-3x}$ (b) $e^{x/3}/3$ (c) $e^{x/4}$ (d) $2e^{2x+3}$ (e) $e^{\sqrt{x}-4}/2x^{1/2}$

8. (a) $1 + \ln x$ (b) $3x + 2x \ln x^3$ (c) $4x^3/(2x - 1) + 6x^2 \ln(2x - 1)$

   (d) $3xe^{3x} + e^{3x}$ (e) $x^3 e^{x^2/2} + 2xe^{x^2/2}$

   (f) $2x^2/(x^2 + 1) + \ln \frac{1}{2}(x^2 + 1)$

9. (a) x/y (b) $(2x + 3)/(1 - 2y)$ (c) $(x - 2y)/(2x - y)$

   (d) $(2x - 9x^2 y)/(4 + 3x^3)$

10. (a) 4   (b) 12   (c) $-12x$ (d) $12x^2 - 24x$

    (e) $-3/4x^{3/2} + 4$ (f) $6x + 6$

11. $6x^2 + 6x - 4$, 32   12. $y = 2 + x^2 - x^3$

13. (a) $y = -7x - 1$, $x - 7y + 43 = 0$ (b) $y = 5x - 2$, $x + 5y - 16 = 0$

    (c) $y = 4x + 8$, $x + 4y + 2 = 0$ (d) $x + 2y - 6 = 0$, $y = 2x - 2$

14. $y = -5x + 10$ 15. $y = -8x - 2$, $y = x + 16$, $y = x - 16$

16. $y = 9x - 16$, $(-4, -52)$

17. (a) 2   (b) $1/2$ (c) $-1/4$ (d) 3   (e) $-4/3$ (f) 4

18. (a) $(-1, -3)$ min.   (b) $(-1, 5)$ max.   (c) $(1, -2)$ min. $(-1, 2)$ max.

    (d) $(2, -16)$ min. $(-2, 16)$ max.   (e) $(2, -8)$ min. $(-2/3, 40/27)$ max.

    (f) $(1/3, -5/27)$ min. $(-1, 1)$ max.

19. (a) (b) (c) (d) (e) (f)

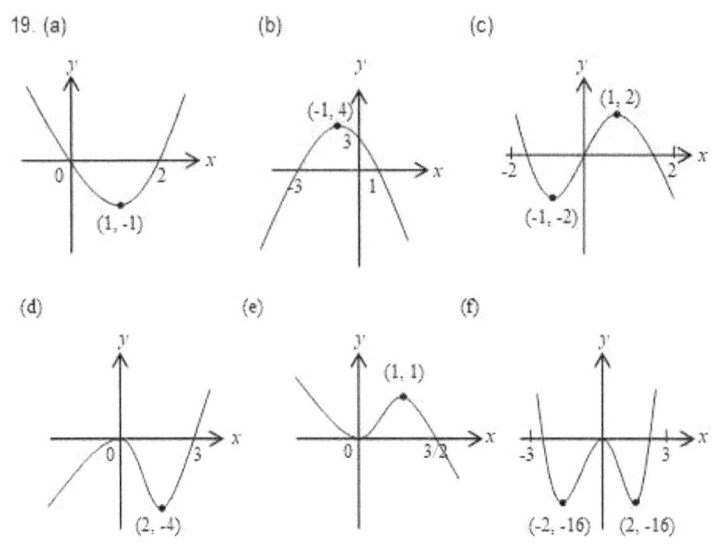

20. 8.6 cm   21. 20   22. 7   23. 60 m, £240

24. 2.7 cm^3 25. 4%   26. $2\frac{1}{2}$%   27. 25 cm^2 s^{-1}

28. 1.24π cm^2 s^{-1} 29. $1/45\pi$ cm s^{-1}

30. (a) $y = x^2 + 1$ (b) $y = x^3 + x^2/2 - 1/2$

   (c) $y = x^3/3 + x/2 + x + 7/6$

   (d) $y = x^4/4 + x^2/2 - 3/4$

31. (a) $x^2 + c$ (b) $x^3 + x^2 + x + c$ (c) $x^3/3 + x^4/2 + 5x + c$

   (d) $3x^4/4 + 7x^2/2 + c$ (e) $x^4/4 - x^3/3 - 3x^2 + c$ (f) $x^4 - 2x^5/5 + c$

32. (a) $\sin 3x/3 + c$ (b) $-\cos 2x/2 + c$ (c) $2\tan 2x + c$

   (d) $-\cos(2x + 1)/2 + c$ (e) $\tan(4x - 3)/4 + c$ (f) $8\sin\left(\frac{1}{2}x + 3\right) + c$

33. (a) $3e^{2x}/2 + c$ (b) $-5e^{-2x}/4 + c$ (c) $4e^{3x}/3 + c$

   (d) $-3e^{-3x+4}/2 + c$ (e) $e^{3x-4} + c$ (f) $2e^{x/2+1} + c$

34. (a) 36   (b) 25/12   (c) 44/3   (d) 57/2   (e) 58/7   (f) 52/5

35. $s = t^3 + 2t^2 + 2$

36. $v = 3t^2 - 8t + c$, $s = t^3 - 4t^2 + 6t + 4$

37. (a) $v = 3t^2/2 - 4t + 6$, $s = t^3/2 - 2t^2 + 6t + 5$

   (b) $v = -10t + 13$, $s = -5t^2 + 13t - 2$

38. $4/3$ m, 1 m s^{-1}

39. (a) $a = 1 - t/6$, $s = 6 + t^2/2 - t^3/36$ (b) 6 s, 18 s

40. (a) 4   (b) 19/20   (c) 6   (d) 1/4

41. 50 m s^{-1} 42. 2 m s^{-2} 43. 40 s

44. 1/5 hr, 187.5 km h^{-2}

Milton Keynes UK
Ingram Content Group UK Ltd.
UKHW041836181124
451360UK00007B/862